The World of Mars

CONTEMPORARY
SCIENCE
PAPERBACKS
35

V. AXEL FIRSOFF M.A., F.R.A.S.

The World of Mars

OLIVER & BOYD *Edinburgh*

OLIVER & BOYD LTD
Tweeddale Court Edinburgh 1

First published 1969

05 001741 1

Set in Times New Roman and printed in
Great Britain by Hazell Watson & Viney Ltd
Aylesbury, Bucks

Preface

Ever since G. V. Schiaparelli announced in 1877 the existence of thin linear markings, known as 'canals', on the surface of Mars, this planet has been attracting both scientific and lay interest as a possible 'abode of life' and civilisation. Being also well placed for telescopic study as our nearest planetary neighbour outwards from the Sun, it is, after the Moon, the most observed and best known of all heavenly bodies. And a fascinating subject of study Mars undoubtedly is – with its waxing and waning polar caps, seasonal changes, clouds and frostings. Yet when in July 1965 the American photographic probe *Mariner 4* radioed back to Earth its 22 close-up frames of the planet, the latter was shown to resemble the Moon rather than the Earth, which came as a shock to most astronomers and poignantly revealed the limitations of telescopic enquiry. The simultaneous estimates of the mass and density of the Martian 'air' from the behaviour of radio signals passed through it by the probe appeared to show it to be much thinner than previously thought, which would make the planet less suitable for life than had been assumed in experiments with simulated Martian environments.

On the other hand, the close-ups of the Moon obtained by the *Lunar Orbiters* have shown beyond reasonable doubt the past existence of temporary rivers on its surface, and Mars as a more massive world would be much better placed in this respect. The atmospheric data derived from the *Mariner* observations involved a sizable element of speculation and are largely incompatible with various well-established findings of telescopic astronomy. Thus the old controversies remain unresolved, and the world of Mars is as mysterious as ever.

In writing the present volume I have endeavoured to keep

all these factors in proper perspective and not to be carried away by the short-lived enthusiasms that tend to sweep across the scientific scene after every new discovery. Fortunately, the subject is manageable within the compass of a paperback, so that nothing of real importance need be omitted from the general picture. The book is intended to be readily accessible to non-astronomers, all technical terms being painstakingly explained for their benefit, without sacrificing the controversial points and novel approaches that may be of interest to the more specialised reader.

Considerable effort has been made to supplement my own modest observational drawings with the best possible illustrations and to keep away from the conventional Mt Wilson and Palomar photographs, excellent as they may be in themselves, which invariably grace the pages of most popular works on Mars. In this connection my sincere gratitude is due to Mr L. F. Ball, Dr C. F. Capen, Dr Audoin Dollfus, Dr Shiro Ebisawa, Mr A. W. Heath, Mr Tsuneo Saheki and Kwasan and Lowell Observatories for supplying many excellent photographs and drawings of the planet, as well as to the U.S. Information Service at the American Embassy in London for the *Mariner 4* frames. My special thanks are due to Mr Takeshi Sato for the information on the Japanese observations of Mars, to the National Physical Laboratory for the data on the vapour pressures of ozone, supplied in a different connection but used also in this volume, and to Dr E. W. Maddison, Librarian of the Royal Astronomical Society, for his unstinted help in the matters of bibliography and reference.

V. A. FIRSOFF

Contents

1. The Red Sky-Wanderer

A *planet*, as distinct from a *fixed star*, is a wandering one – from the Greek *planan*, meaning 'to lead astray'. As a matter of fact, even a fixed star is not pinned fast to the vault of heaven as the Ancients used to think: it has a small but astronomically measurable proper motion, and very slowly shifts position among its celestial neighbours. It will take, however, hundreds and even thousands of years for such shifts of the nearest stars to become noticeable to the naked eye. This is how the Alexandrian astronomer Ptolemy, who compiled his 'Great Array of Heavens', popularly known under the Arabic name of *Almagest*, in about 130 B.C., was able to get away with his 'sphere of fixed stars', where they were supposed to preserve an eternal pattern.

The planets, however, led the early astronomers a bewildering dance among the stars before returning to the same, or only approximately the same, spot in the sky. Had the celestial clock been made in Switzerland it would, no doubt, have jewel bearings and all its cog-wheels nicely round and centred; but it 'just growed', and Ptolemy, who with the rest of the Ancient World believed that the 'divine harmony of the spheres' required all planetary motions to be in perfect circles, had great difficulty in fitting the observations into this system. The Earth, of course, was the centre of the universe, as any person of good sense could see at a glance. Nor could it be spinning, as some Pythagoreans proposed, or else all objects unattached to its surface would be flung off into space like mud off the rim of a revolving wheel – this stands to reason, or so Ptolemy thought, having no inkling of gravitation.

Thus the crystal 'spheres' turned round the Earth with

sweet music, inaudible to mortal ears, carrying with them the 'planets' in ever-widening circles to the mystical number of seven: Moon, Venus, Mercury, Sun, Mars, Jupiter, Saturn, followed by the 'sphere of fixed stars' as an ornamental back-cloth.

Yet, to take Mars for an example, the Red Wanderer, dedicated to the God of War from its colour, behaved most erratically. The speed of its celestial progress varied, and at a certain point it stopped, then doubled back on its tracks, to halt again and resume a westward advance. Clearly, thought Ptolemy, following a suggestion made by Apollonius of Perga, Mars must be fixed to another sphere, which was mounted on the perimeter of the main one and revolved on its own. The two circular motions became known as the *deferent* and the *epicycle* respectively, and their combination was called into account for the effect. As the accuracy of measurements increased (there were no telescopes until A.D. 1609), however, more and more epicycles had to be added to keep track of the elusive planet, until the system of spheres began to look like a bunch of soap bubbles of decreasing diameter, growing progressively more and more unwieldy.

This was the kind of astronomy taught in the mediaeval *quadrivium*, where it shared honours with arithmetic, geometry and music, as was only fitting.

Nevertheless, some 400 years before Ptolemy's time there was born in Samos a child by the name of Aristarchus, who was very bright at geometry, and on growing up proceeded to measure the size and distance of the Sun from the triangle drawn from the observer to the centre of the Moon at *dichotomy* (exact half-phase) and that of the Sun. The method was foolproof in theory, but difficult to apply in practice, and the figures as determined by Aristarchus came out some 21 times too low, as we now know. He put the Sun 4 800 000 miles away and made it 40 000 miles across. The diameter of the Earth had already been determined quite accurately at 7900 miles by Eratosthenes, and Aristarchus concluded that the Sun was much the larger and more important of the two. He therefore put it at the centre of the system (and of the universe), with Mercury, Venus, the Earth, Mars, Jupiter and Saturn revolving round it, although the Moon continued to circle the Earth, while the diurnal motion of the heavens was due simply to the spin of the Earth.

This is very nearly what we think today, but was far in advance of the time, for it was, to say the least, heretical to equate the solid globe of the Earth (the Flat-Earth Society was already out of date), the abode of man- and god-kind, with mere wandering lights, such as Mars or Venus. Furthermore, Aristarchus continued to rely on perfect circles, and in these terms his system was mathematically unsatisfactory. The great Ptolemy made short shrift of this amateurish effort and the 'scientific establishment' followed suit for 1600 years, until, eventually, Copernicus reverted with great trepidation to Aristarchus' point of view on the basis of more accurate observational data.

Copernicus determined with fair accuracy that the orbital radii of the Earth, Mars, Jupiter and Saturn (the trans-Saturnian planets continued to await discovery for another 300 years or so) were in the ratio of 1:1·5:5·2:9, but his value for the Earth's mean distance from the Sun, known as the *astronomical unit*, or *A. U.* for short, was still that found by Aristarchus. (The latest figure, based on radar echoes, is 149 000 000 km = 92 957 209 miles.)

He was also a believer in the perfect circle, and his daring extended no further than placing the Sun eccentrically within it. The resulting system was not wholly satisfactory, though it marked a great intellectual revolution, until the momentous year of 1609, when Kepler formulated his laws of elliptical planetary motion, thus reviving the forgotten 600-year-old idea of the Arabic astronomer Arzachel (Allah was a less exacting geometer than Zeus). Everything began to fall nicely into place. Came Newton with universal gravitation to explain things and 'all was light', as bright as hindsight and textbooks can make it, the darkness having retreated into less accessible corners of the field where it still dwells.

Today the Ptolemaic system of the world lingers thinly in epicyclic gearing with its planet-carriers for deferents, but even this is heliocentric, as it has central sun-wheels. *Sic transit musica sphaerarum*, as instrumented probes speed on Keplerian orbits from world to world on the waves of Newtonian gravitational flood. Yet it is salutary to reflect how recent are these historical developments and how fierce the opposition to every advance has been.

Counting the planets from the Sun, the Earth is the third out, Mars the fourth. The *mean semi-diameters* of their orbital ellipses are 1·000 A.U. (by definition) and 1·524 A.U. (from observation) respectively. The deviation of an orbit from the circle may be expressed as its *ellipticity*

$$\varepsilon = \frac{a - b}{a},$$

where a stands for the half of the major axis $2a$, and b for the half of the minor axis $2b$. It is more usual, however, to define this quantity by the related concept of *eccentricity* e, given by the Pythagorean relationship:

$$e^2 = \frac{a^2 - b^2}{a^2}.$$

With an eccentricity of 0·0167, the Earth's orbit is nearly circular, but that of Mars has an eccentricity of 0·0934 and is noticeably stretched. The Sun lies at one of the foci of the ellipse, so that Mars at the point of its orbit closest to the Sun, which is called *perihelion*, is 128 million miles from it, and at the farthest point, or *aphelion*, this distance rises to 154 100 000 miles, which has manifold consequences. The plane of Mars's orbit is inclined to that of the Earth's circumsolar path at the present angle of 1°50′ and 59·6″ – for orbital elements, too, are subject to slow secular change.

Orbital velocity must be such as to counterbalance the gravitational attraction of the Sun's great mass, which exceeds that of the Earth 333 000 times. Thus, like the Red Queen, the Earth must move very fast to keep its orbital station, its mean velocity being 18·52 miles per second. This velocity, however, is inversely proportional to the square root of the orbital semi-diameter, and is correspondingly lower for Mars, which averages 15·01 miles per second, being faster at perihelion and slower at aphelion. From this it follows that the Earth, running faster along a smaller circuit, will be overtaking Mars on its slower and longer journey.

The jay-walking of Mars about the sky is the result of superimposition of these two orbital movements in projection upon it. The two ellipses being inclined to each other and the orbital velocities of both planets varying with the distance from the Sun, it is clear why Mars does not return to the same geometrical configuration with the Earth and the Sun after a fixed period. The mean value of this *synodic period* is 780 days, or if you are a stickler for precision 779·94 days.

The regression of Mars and of the other *superior* or *exterior planets*, which move beyond the Earth as seen from the Sun, is due to the combination of the Earth's and the planet's movements, as explained in Fig. 1.

If the plane of the Earth's orbit is produced to intersection with the apparent sphere of heaven it will define a great circle along which the Sun rolls over the year and where eclipses occur, whence it has been given the name of *ecliptic.* The orbits of major planets (though not necessarily those of minor planets or planetoids) do not generally deviate much from the plane of the ecliptic and intersect it at small angles. As we have seen, this angle in the case of Mars is just short of 1°51′, so that Mars may rise above or drop below the ecliptic by a corresponding arc.

The ecliptical motion of the Sun mirrors the orbital progress of the Earth, but the risings and settings of celestial bodies, as well as their *culminations*, when they cross the local meridian and stand highest in the sky, reflect the *axial spin* of the Earth and the geographical location of the observer. If the Earth's polar axis stood upright to the plane of the ecliptic, the latter would coincide with the *celestial equator*, which is another great circle at the intersection of the equatorial plane with the celestial sphere, night and day would be of equal length, and there would be no seasons, which might make the British Isles uncomfortably chilly. In reality, however, the polar axis is tilted 23°27′ away from the normal to the ecliptical plane which, consequently, crosses the equatorial plane at this angle.

The Sun attains its highest point of the sky at summer

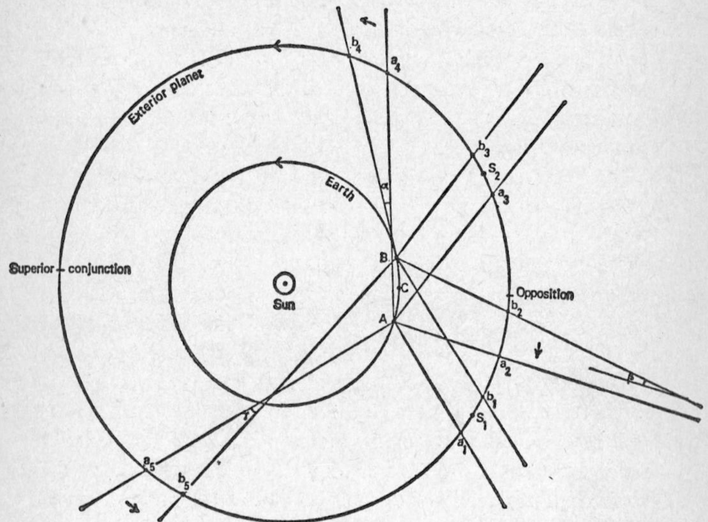

Fig. 1. *The apparent movement of an exterior planet in the sky. For the sake of simplicity the orbits are assumed to be circular and coplanar with the ratio of radii of 1:2. The marks for opposition and superior conjunction correspond to the position of the Earth at* C. *Both planets move counter-clockwise, the Earth* $\sqrt{2}$ *faster than the planet. In a given time interval,* t, *the Earth will move from* A *to* B *and the planet from* a_n *to* b_n. *Five different positions of the planet have been illustrated. In moving from* a_5 *to* b_5, *the planet as seen from the Earth and indicated by the sighting lines* A–a_5 *and* B–b_5 *will have moved forward (in the sense of revolution) in the sky through an angle* γ. *But the sighting lines* A–a_1 *and* B–b_1 *are parallel and meet at infinity at the same point of the sky, so that there is no apparent movement, or to be exact, the planet has moved by the same angle forward and backwards, the reversal of sense of motion occurring at the stationary point* S_1. S_2 *is the second stationary point, where the sense of motion is reversed once more. Between* S_1 *and* S_2 *the planet retrogrades. Thus between* a_2 *and* b_2 *the planet moves back through an angle* β. *Beyond* S_2, *however, its apparent motion is direct, as shown at* a_3 *and* b_3.

solstice, when it stands 23°27′ above the celestial equator, and its lowest point at winter solstice, when it climbs down below the equator by the same amount. It follows that during the summer the half of the ecliptical circle containing the Sun rises above the celestial equator, while the night half dips below it. The planets, however, which stay close to the ecliptic, are generally best seen by night, so that they will keep to the comparatively low altitudes in summer. In winter the situation is reversed.

Now, the air is an inevitable and the most uncertain part of the telescopic optics. It contains obscurations and tremors which can make the image unsteady, causing what is known as 'boiling' in astronomical parlance. If the whole of the atmosphere directly above us at sea-level were compressed to sea-level density at freezing point it would make a layer about 5½ miles deep. This is known as the *equivalent atmosphere*, a planetologically important concept, or an *air mass*. Thus looking at a celestial body at the zenith from the seashore we will have one air mass in the line of sight, but atmospheric mass in the line of sight increases as the secant of the zenithal angle. If, therefore, that body is 45° above the horizon there will be 1·41 (or $\sqrt{2}$) air masses between it and the observer, and 2 air masses if its altitude declines to 30°. Atmospheric disturbance will increase at the same rate, and as a rule a planet below 30° from the horizon is hardly worth looking at.

An exterior planet, such as Mars, moves completely round the celestial sphere each synodic period. It is most distant from the Earth at *superior conjunction*, when the Earth, the Sun and the planet lie in line with the Sun in between the two. The actual distance will vary owing to orbital eccentricities and inclinations, but the maximum for Mars is 340 million miles. Conversely, the closest approach of the planet to the Earth will occur at *opposition*, when the three bodies are aligned with the Earth and the planet on the same side of the Sun.

At opposition the planet is exactly opposite the Sun on the celestial sphere, or 180° from it either way. The planet will, accordingly, rise at sunset, culminate at local midnight and set at sunrise, being visible all night long.

It does not require much imagination to grasp why an opposition represents the best possible configuration from the observer's point of view. Its only limitation is that the planet shows a full disc, and a full disc is devoid of shadows cast by surface relief, which becomes difficult to spot.

But not all oppositions are equal by reason of orbital eccentricities, the changing position of the planet in its orbit relatively to the ecliptic and the point of the ecliptic corresponding to the opposition. The oppositions of Mars follow one another after a synodic period of approximately 780 days,

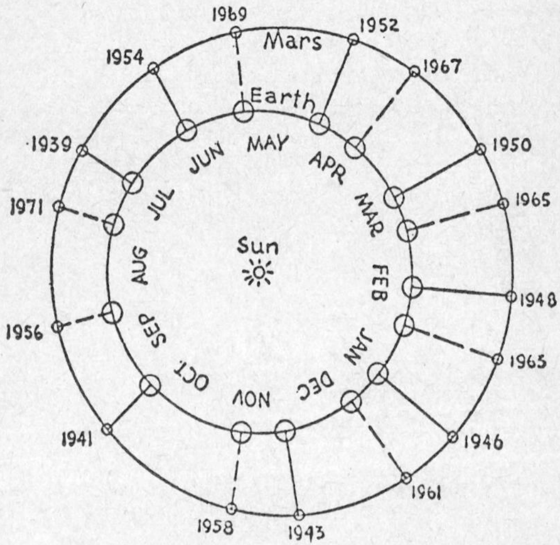

Fig. 2. *Oppositions of Mars, 1939–71.* (*From* Earth, Moon and Planets *by F. L. Whipple, and reproduced by courtesy of Harvard University Press.*)

which is not a multiple of the year, so that they move round the seasons, as well as round the orbit of Mars, whose 'year', or *sidereal period of orbital revolution*, is equal to 686·979 Earth days (Fig. 2).

As a result, the distance of Mars at opposition varies from 34 600 000 to 62 900 000 miles. Winter oppositions are unfavourable, but for the reasons explained above Mars rises high in our skies. A midsummer opposition brings it closer, but it stays uncomfortably low for observation from high northern altitudes. The best possible combination of distance and visibility occurs in late summer or early autumn.

At the moment we are passing through a period of unfavourable oppositions, but on 6 August 1971 Mars will make its closest possible approach of 34 600 000 miles to Earth and will be excellently placed for observation, especially from the southern hemisphere.

With large remote worlds, such as Jupiter or Saturn, the maximum variation in distance as between an opposition and a superior conjunction is a matter of about 2 A.U., which has little effect on their apparent diameter. But Mars is small, with a diameter of 4220 miles and is the closest planet after

Fig. 3. *Scale comparison of the apparent size of Mars at favourable opposition, unfavourable opposition and superior conjunction.*

Venus. Its distance may change from 340 000 000 miles to 34 600 000, or nearly tenfold. Since the latter corresponds to an apparent diameter of 24·9″, this will shrink to about 2·5″ when the planet is farthest at superior conjunction, and a disc as small as that shows very little detail even through the largest telescopes (Fig. 3).

The difference of diameter as between a favourable and an unfavourable opposition is not inconsiderable either, and at the most unfavourable approach Mars is no larger than 13·5″. The brightness follows suit.

It is expressed rather awkwardly in terms of magnitude, an object one magnitude *lower* than another being 2·512 times *brighter* than it. The only convenient point about this scale is that log 2·512 = 0·4000. Thus when closest, Mars of magnitude −2·8 outshines Jupiter and is second only to Venus among the 'stars' of the firmament. But at a least favourable opposition the magnitude of Mars rises to −1·1, which places it below Mercury and Sirius, the brightest of the fixed stars.

Be this as it may, the really useful telescopic study of Mars is limited to a few weeks about the date of an opposition, and owing to the general interest in the Red Planet its oppositions have become great astronomical events. Special international arrangements, known as the 'Mars Patrol', are made to keep watch on the planet round the clock and round the world. Since Mars rotates on its axis in about the same period as the Earth, this allows to cover the entire globe and not to miss any important happenings, such as, say, a volcanic eruption or an unusual cloud formation.

2. A Little Earth

Mars has been dubbed '*the* Little Earth', with, it seems, more imagination than accuracy. But it certainly is *a* little Earth, which is another way of saying that it is a small Earth-like or *terrestrial* planet.

There are definitely four, perhaps five, such planets in the Solar System. They are characterised by small dimensions and masses, solid rocky surfaces and moderate or thin oxidising atmospheres. Mercury, Venus, Earth and Mars fall into this category, but does Pluto? It is often forced into it in astronomical tables, but the observational data taken in the raw would give it approximately the size of Mars with a mass of Venus, which results in a mean density of over 20 and a surface gravity 2½ times higher than on the Earth (1 *g*). Is this still a terrestrial planet? It would be a world very different from ours, even quite apart from its great distance from the Sun, which should make it very cold (an estimated mean annual temperature of 59°K = −214°C). Those who profess that so high a density is improbable may well be reminded of dwarf stars, composed of degenerate matter, which comprise the mass of a Sun in a volume no larger than Pluto's, while the equally massive, if so far hypothetical, neutron stars may be only about the size of the Moon.

On the other hand, if, say, the biggest moons of Saturn and Jupiter, Titan and Ganymede, equalling or exceeding Mercury in diameter, pursued independent circumsolar orbits, they would have to be classed, in spite of their smaller masses, as terrestrial planets. Indeed, Ganymede bears a certain resemblance to Mars, and has even been credited, like it, with polar caps, which would presumably consist of solid ammonia.

Finally, would a planet three or four times as massive as the Earth still be a terrestrial? There must be such planets somewhere among the limitless universe of stars.

All this amounts to saying that the term *terrestrial planet* does not really mean very much unless applied to the four *inner planets* of the Sun.

Mars is the outermost of them, 1·54 A.U. away from this source of life-giving energy. Beyond it extends a great gap without a major planet, which is filled thickly between 2·7 and 3·2 A.U. and thinly outside these limits by a swarm of minor planets, planetoids or asteroids, ranging from a few to a few hundred miles in diameter, and then at 5·20 A.U. comes Jupiter, a huge semi-stellar body, 88 700 miles across and 318 times the Earth's mass. It is this situation that differentiates so clearly the inner terrestrial planets into a distinct solar sub-family.

Distance from the Sun is very important for several reasons, the foremost of which is the intensity of sunshine that is inversely proportional to its square. Terrestrial planets have insufficient internal heat to affect their surface conditions, which depend primarily on the energy of sunlight.

Thus the radioactive heat of the Earth is leaking out from the interior at the rate of 0·000 072 small calories per square centimetre of the surface per minute; whereas the *solar constant*, measuring the mean radiation of the Sun at the Earth's orbit, is 1·99 (say 2) cal/cm^2/min. This radiation is intercepted by the Earth and other planets in proportion to their cross-sectional area. The internal heat, on the other hand, is reaching the entire surface of the globe, so that to compare the two it must be multiplied into four, which is the ratio of the superficies to the cross-section of a sphere. This will raise the contribution from inside the Earth to 0·000 288 cal/cm^2/min – still a negligible fraction of the inflow of solar energy.

At the distance of Mars the solar constant is pared down to 0·43 of its terrestrial value, but this is an insignificant reduction as compared with any likely evolution of heat from the planet's interior. For, although minor differences in the chemical make-up of terrestrial planets must be expected, they have a common origin and mean densities of between four and six times that of water, so that there can be no great qualitative disparity in their materials. Even quantitatively their mineralogies would be closely comparable, and this goes also for the radioactive elements, to which the internal heat of the Earth is ascribed. The case of the large jovian planets is rather different, but does not concern us here.

What little is known about the birth and evolution of stars makes it most plausible that the Sun and its family of planets arose from the same primordial cloud, or *galactic globule*. As the cloud condensed under the action of its own gravitational attraction, external pressure of radiation and inner rotational friction, something resembling cometary bodies began to form within it, and the planets are thought to have grown gradually by picking them up. This matter was cold to begin with, for had it been hot the Earth could not have acquired either air or water. But the main mass of the nebular material was concentrated in the densest inner portion of the cloud, and here the temperature would be rising rapidly by gravitational contraction alone, until the central mass began to shine – the Sun was born.

Its radiation heated up and ionised the surrounding cloud of gas, dust and 'ices', now contracted into a disc, driving its lighter, more volatile constituents away towards the periphery, as it still is the matter of cometary tails. This is why the small terrestrial planets, composed of relatively heavy matter, which is cosmically scarce, have been formed inside the 'inner circle' as defined by the orbit of Mars, while the massive jovian globes have evolved beyond 5 A.U., where hydrogen and helium, which between them accounted for most of the nebula, were readily available.

Yet a similar differentiation according to specific gravity must have operated within the 'inner circle' as well, its relative effectiveness depending on the timing of the process of planetary accretion in relation to the emergence of the Sun into full stardom, ionisation and chemical bonding intervening further to complicate matters. Taking one consideration with another, it seems that the planets were already there, at least in embryo, when the Sun blazed out, which may have happened quite suddenly on reaching the critical condition. Anyway, on a photograph of an area of the sky in Orion taken in 1954 were found two stellar points, now known as 'Herbig's Objects', which were not there when the same area was photographed in 1947. Thus a period of seven years appears to span the moments of birth of two stars.

As Galileo rightly perceived, the satellite system of Jupiter is as though a solar system in miniature. I seem to have been the first to draw attention to the fact that the product of the mean density of Jupiter's large Galilean moons into their distance from the planet's centre is approximately constant; in other words, density is inversely proportional to distance. Jupiter must have been a star when the satellites were formed.

In the Solar System the relationship is not quite so simple, most probably owing to the much greater differences in the masses of the planets, which affect both the surface gravities and so the ability to retain gases on the outside and the gravitational compression inside, thus concealing the true progression. Nevertheless, a general decline in density from Mercury to Saturn is unmistakable. Beyond Saturn mean densities begin to rise again, which must be ascribed to the progressive escape of hydrogen to space from the weakening gravitational grip of the Sun or its natal nebula.

Mercury, a body with a mass about a half that of Mars, is credited with mean densities between 5·4 and 6·2, depending on the assumed diameter. The degree of internal compression being comparatively slight, this makes Mercury absolutely the densest of the terrestrials. Equally there can be no doubt that Mars, having the average specific gravity of 3·94, is the least dense of them and so contains a higher proportion of lighter and a lower proportion of heavier atoms in its chemical make-up than does the Earth.

This may have various geological implications.

We have, of course, no direct way of examining the interior not only of Mars, but even of the Earth, and our knowledge of these is purely inferential. It derives mainly from two sources: the composition of meteorites, taken to exemplify the material of planetary interiors, and the behaviour of earthquake (seismic) waves passing through the Earth. Our deepest bore-holes reach no farther down than 5 miles, which is a mere pinprick to a body 3960 miles in radius, but the paths and velocities of transverse and compressional earthquake waves have been extensively studied.

These have shown that the globe of our planet is stratified in layers differing in density and mechanical properties (Fig. 4). The outermost skin, of the average continental

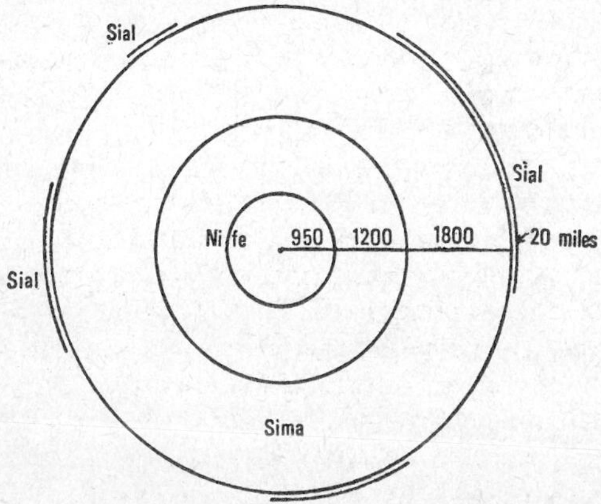

Fig. 4. *Internal constitution of the Earth. The distribution of sial is patchy and cannot be shown accurately on this scale.*

thickness of 20 miles and largely absent from deep-ocean beds, consists mainly of acid igneous rocks with a high proportion of silica and the sedimentary products of their decay. It is called *sial* from the predominance of silicon and aluminium. This is the crust, which is the main seat of the geological processes and the substratum of life. Geology, however, cannot afford to overlook the underlying *mantle* of the Earth, which forms the semi-plastic support for the crust, now subsiding under its increasing weight, now redressing itself when the burden has been lightened by erosion, in up-

and-down movements of *isostatic* adjustment. Isostasy and crustal shortening also preside over the elevation of alpine mountain ranges, faulting and thrusting.

The mantle is believed to be composed of heavy basic rocks, typified by dunite and peridotite and some stony meteorites, with a preponderance of silicon and magnesium in its chemical make-up, whence it has been given the name of *sima*. It extends to a depth of 1800 miles below the sialic crust.

The average specific gravity of surface formations is actually only about 2·3, but the mean for the sial layer is 2·70, there being a sudden jump to 3·32 when the top of the mantle (Moho) is reached. Within the mantle, the density rises gradually, chiefly by compression mediating changes of crystalline habit, on the way down, to become 5·68 at the base, where a sharp increase to 9·43 marks another internal discontinuity. This is ascribed to a change of composition, the heavy *core* resembling iron-nickel meteorites and being referred to as *nife* from the chemical symbols for nickel (Ni) and iron (Fe). It behaves as a liquid and is 1200 miles deep. 950 miles from the Earth's centre, however, there is one more discontinuity, with a density jump from 11·54 to 16·8 and a change in behaviour to that of a rigid solid. The condition and composition of this *inner core* are conjectural, but temperatures of the order of 10 000°C are expected, so that it must be gas.

Having originated from the same galactic globule astronomically close to the Earth, Mars will have, as stated on page 13, a similar overall composition. Yet its lower density of 3·94 indicates, after an allowance has been made for reduced internal compression, an expectedly higher proportion of lightweight materials, including some substances relatively scarce on the Earth, with a corresponding reduction in the heavier elements. In other words – more sial and less nife.

Something can be learned about the internal structure of a planet from its polar flattening, or *oblateness*, which is the result of interaction between gravitation and axial rotation. The condition of gravitational equilibrium is achieved in a perfect sphere, but the centrifugal forces generated by the spin, which are strongest at the equator and peter out towards the poles, cause a rotating body to bulge out at the former and to sag in at the latter. The oblateness, or *ellipticity*, e, is measured by the difference between the equatorial radius R_e and the polar radius R_p divided into R_e; in other words

$$e = \frac{R_e - R_p}{R_e}.$$

Since the centrifugal drag is directly proportional to the mass and its distance from the axis of revolution, it is clear that the faster the planet spins, the larger it is and the less its mass is concentrated towards the centre, the more oblate, or flattened at the poles, it must be.

There exist comparatively simple mathematical equations relating these factors to one another and oblateness to the internal distribution of mass.

Now the equatorial radii of the Earth and Mars are 3963 and 2110 miles respectively. The *sidereal day*, between two successive transits of the same star observed from a point on the surface, is 23 hours 56 minutes and 4 seconds in the first case, and 24 h 37 min and 23 s in the second. Thus the centrifugal forces within the body of Mars are smaller, yet its oblateness of 0·0052 is higher than the Earth's 0·0034. Thence it is inferred that the mass of Mars is less concentrated towards the centre. To some extent this is the inevitable consequence of its smaller mass, which is only 0·107 of our planet's, for the central density of the Earth, put at 17·2, is due to compression, arising from increased gravity acting on a greater thickness of superjacent rocks. Nevertheless, the oblateness of Mars forms a close approach to the theoretical model of a homogeneous planet of uniform density throughout.

In the model constructed by Sir Harold Jeffreys, Mars is credited with a small iron-nickel core, but it has been mooted that it may have no core at all, there having been no gravitational fractionation of material, which requires the interior to have passed through a liquid phase. This, however, does not appear very probable.

Taking the terrestrial and meteoritic abundances of radioactive elements as the starting point, G. P. Kuiper has calculated that 4500 million years ago, when the planets were formed, the heat released by the decay of these elements will have been ten times higher than at present, and concluded that all bodies over 100 miles in diameter must have melted inside as a result. It could, of course, be that Mars was spinning faster to begin with and acquired its present figure at that time, the excessive oblateness having been preserved into the present stage of its evolution by the hardening of the light surface layers. Sial has a much higher melting point than sima. The Moon, too, is excessively flattened at the poles, which calls for a similar explanation.

This is substantially the conclusion reached by D. L. Lamar, who investigated the situation in 1961, without, however, invoking the evolutionary explanation just mentioned. He distinguishes between the observed *optical* oblateness and the *dynamical* oblateness of the invisible equipotential surface over which the forces of gravitational pressure and centrifugal drag are isostatically balanced. Assuming that the crust of Mars is 0·5 gm/cm^3 lighter than the mantle, he finds that the discrepancy can be removed by variation in crustal thickness alone, this being between 18 and 175 km, with the isostatic equilibrium attained at depths of over 250 km, say 150 miles.

Too much importance must not be attached to such numerical results, but in agreement Jeffreys' reasoning these indicate a thick sialic layer. In other words, basic rocks cannot be expected to outcrop on Mars and its surface formations will be highly acid and include many light mineral species.

By the same token Mars must have received at birth a proportional allowance of water, ammonia, methane and other gases exceeding that of the Earth. Since these substances were acquired by the aggregation of cold cometary and meteoric matter, the outcome could not have been seriously affected by the lower mass of Mars; and since the primitive atmospheres of the terrestrial planets have been lost anyway during the later hot-surface stage (see *The Interior Planets* in the same series for fuller discussion), only such volatile constituents as have been chemically bound or occluded in the interior could have contributed to the present situation in either case. The remaining secondary atmosphere has been produced by volcanic exhalation.

Yet Mars has no seas and only a thin atmosphere. The latter is cold. The surface gravity of 0·38 *g* and the escape velocity of 3·1 miles per second are wholly adequate for retaining most atmospheric gases, excepting hydrogen and helium, and water for cosmic periods. The loss of surface water through photo-dissociation by ultra-violet radiation at high atmospheric levels is one of the common clichés without much substance, as the ascending water vapour will be entirely frozen out at the cold trap. Thus the question poses itself, What has become of the volatiles originally trapped inside Mars?

Here Mars invites once more comparison with the Moon rather than the Earth, for the Moon presents a similar problem.

It is true that a body of small mass, having a low escape velocity, will be steadily losing some of its outer gas envelope by molecular evaporation to space, as well as by the sweeping action of the solar wind of fast protons emitted by flares. It seems, however, that the latter effect has been exaggerated. Moreover, it is not all loss, for there will also be gravitational and collisional capture of interplanetary gas (Firsoff, *Science*, 1959), whose density rises with the solar wind. In Jeans' original investigation (1925) of the problem of dissipation of planetary atmospheres an atmosphere of a given gas will endure for over 1000 million years (one acon) if its mean-square velocity is less than a fifth of the velocity of escape, and we have seen that Mars ought to satisfy this condition for most gases. But the problem is immensely complicated and no simple mathematical treatment can give a true picture of the situation. Jeans considered the atmosphere to be isothermal throughout. In actual fact, however, temperature drops steadily with height in the lower *convective* atmosphere, or *troposphere*, which contains most of the atmospheric mass. The overlying *stratosphere* may fit his model in part, but there, too, the temperature varies at higher levels owing to the presence of such gases as ozone, which becomes heated by absorbing ultra-violet radiation, and to ionisation. The existence of cold traps may substantially remove some atmospheric constituents from the escape level, so that they will be preserved in defiance of such simple theory. On the other hand, heavy gases may be lost through photo-dissociation into lighter components.

Moreover, neither Jeans himself nor those who followed in his footsteps paid any attention to the hydrosphere. This point was taken up by J. J. Gilvarry in 1960 for the case of the Moon. Starting from the reasonable assumption that the original lunar magmas contained the same proportion of water as their terrestrial counterparts, he found that the Moon ought to have preserved a substantial hydrosphere for at least one aeon, allowing life to develop, and the maria of its hey-day would have been true seas with an average depth of 1¼ miles.

This sounds exciting, but, unfortunately, there is little indication of massive water action on the surface of the Moon, with some highly localised exceptions, such as temporary watercourses. Our knowledge of the Martian surface is too scanty for such an assertion, but the available evidence points to a very similar general situation, although Mars, being some eight times as massive as the Moon, Gilvarry's reasoning should apply to it all the more.

A little digression seems not out of place here. There exists a general tendency to think of water and oxygen, which are vital to animal life known to us, in magical rather than scientific terms. Their presence on any other world is always regarded as something extremely improbable and requiring special proof. Yet as a matter of sober fact hydrogen is by far the commonest of all chemical species in the universe and the next commonest, passing the chemically-inactive helium and neon, is oxygen, so that it is the absence of water that should be a matter for comment, while at least some oxygen must be formed by its dissociation.

It is, therefore, the apparent scarcity of water on the surface of Mars, which will be considered later on from the observational and ecological standpoint, that requires explanation.

One point to be borne in mind is that, while it may be mathematically convenient to assume that a planetary atmosphere was supplied all ready-made in a single fiat, this is quite obviously untrue, for it was produced over a long 'geological' period by volcanic action, which has not ceased yet withal. Instead of surging to the surface of the Moon or Mars in a great flood, the *juvenile* waters were arriving there in little driblets, with perhaps an occasional 'pipe-burst'. Thus in considering a planetary atmo- or hydrosphere we must think of it in the terms of a current account, with small gains and losses to be added and deducted all the time, rather than of a single big win on the pools that was subsequently squandered.

I suspect, however, that even this obvious fact does not provide the complete answer. For it applies to the Earth as well, and if the direct effect of reduced gravity as such is insufficient to account for the observed differences, it is necessary to explore its further consequences, as well as the thermal factors.

Low gravity, combined with the absence of surface water and its consolidating action, will cause surface rocks to be lightly compacted and the primary or igneous formations to assume the form of bubbly (vesicular) lava. Such rocks will be highly absorbent of gas, including water vapour, and so tend to suck up and deplete the atmosphere. They may also be specifically lighter than water (Firsoff, 1959), which would then underlie them in the order of natural gravitational statification, not appearing on the surface in ordinary circumstances. Moreover, for different reasons, the subsoil of the Moon and Mars will be permanently frozen at a certain depth, thus forming a *permafrost seal* (Firsoff, 1959; T. Gold, 1960), so that the water rising to the surface from the interior will be caught in it and frozen on the way, with the exception of geysers or volcanic vents where the seal is broken.

It thus seems that in Martian conditions most of the water will stay locked underground, so that Mars has never had any seas. This has important and varied implications, one of which is a further lowering of the mean density of the crust.

The time is now ripe to examine Mars in the light of observational data.

3. At the Eyepiece

'In colour largely lies this awakening touch that imbues the picture with the sense of actuality. And very vivid are the tints, so salient and so unlike that their naming in words conveys scant idea of their concord to the eye. Rose ochre dominates the lighter regions, while the robin's-egg blue colours the darker; and both are set off and emphasised by the icy whiteness of the caps. Nor is either hue uniform; tone relieves tint to a further heightening of effect. In some parts of the light expanses the ochre prevails alone; in others the rose deepens to a brick-red, suffusing the surface with the glow of a warm, late afternoon. No less various is the blue, now sinking into deeps of shading, now lightening into faint washes that in places grade off insensibly into ochre itself, thus marking regions of intermediate tint the precise borders of which are not decipherable by the eye.' Thus wrote Percival Lowell, who was a great enthusiast and telescopic student of Mars, in *Mars as the Abode of Life* (p. 74), published in 1909, when much excitement and speculation was being caused by the 'canals', first described by Schiaparelli during the favourable opposition of 1877, and the Martians suspected of being responsible for their construction. Lowell held very definite views about both, and we shall have occasion to consider these briefly later on. But his description of the telescopic appearance of Mars is substantially correct, regardless of any hypotheses, although there may be some disagreement about the exact colorations. Thus many observers would describe the dark areas, which have become known as *maria*, the Latin for seas, although they are definitely not water, as grey or grey-green rather than blue, and some of the colour effect is undoubtedly due to the contrast with the predominantly yellowish and reddish tone of the bright *terrae* (lands) (Fig. 5).

The Russian astronomer N. A. Kozyrev, after observing Mars with the 50-inch reflector of the Crimean Astrophysical Observatory during the 1954 opposition, decided that the colourings seen on Mars were illusory and due solely to the

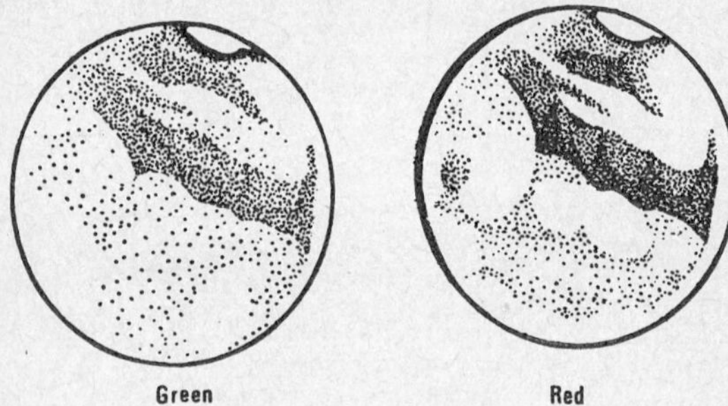

Fig. 5. *Evidence for the green colour in the maria. Observational drawings, by the author, of Maria Sirenum and Australe (part) in green and red. The maria show dark structure in the red; all detail is enhanced, except for the polar collar, which is darker in green. A slight phase effect is seen in red only. 20 August 1956: 2 h 10 min U.T. (red); 2 h 30 min U.T. (green). Aperture 6½ inches, 300 X. Direct estimate of colours: disc – orange; maria – bluish-green; polar cap – sandy-white.*

properties of the Martian atmosphere, which was partly opaque to the light of wavelengths shorter than 5000 Å, this 'green haze' making all bright areas appear reddish and dark ones greenish, though in reality they were all of the same colour, differing in shade alone. This view may contain a certain measure of truth. Our own air is blue and, when superimposed on a dark planetary marking, will impart to it a bluish hue; a bright red area will form a green after-image in the eye, giving rise to a spurious green effect. Yet such explanations cannot account for colour differences that appear simultaneously or periodically on the Martian surface.

Not all maria and terrae, which are often referred to as

and probably are 'deserts', look alike. Thus the prominent dark wedge of Syrtis Major is usually described as blue, while, say, Mare Sirenum appears green. Some terrae are of a pale sandy or even lemon yellow, while others look distinctly reddish at the same time.

In the early comparisons of Mars made at the eyepiece with various terrestrial materials or paints by W. H. Pickering and P. Lowell himself in the U.S.A. the redness of the terrae was greatly exaggerated, with such colours as 'dragon's blood' predominating. Mars emerged from these looking rather like the Painted Desert of Arizona, which may be not unconnected with the location of Lowell Observatory. G. P. Kuiper, however, has found the bright lands of Mars no redder than brown felsite (an acid volcanic rock) in his analysis of the planet's reflection curve.

Most spurious colour side-effects can be weeded out by the use of appropriate filters, whether photographic or visual. Depending on the width of the 'cut', a colour filter transmits a wider or narrower band of wavelengths only and gives a monochromatic image, where all other colours have been eliminated and appear by default as darkenings. In a tri-colour separation set of filters each filter reproduces more or less accurately one of the three basic colour responses of the human eye (red, green and blue): between them these three responses yield all the colours of the rainbow, and any observable colour effect can be split by such a set of filters into its constituent elements and correctly estimated (Fig. 6).

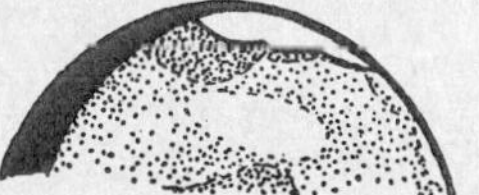
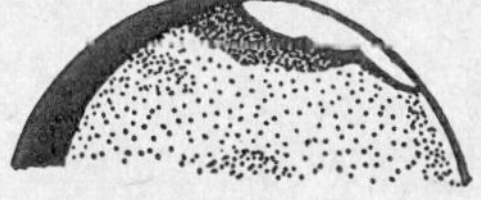

Fig. 6. *The southern part of Mars between Mare Erythraeum and the Pole, from observational drawings by the author on 23 July 1956, 2.30 U.T. The* left *view is through a red filter, the* right *through a green filter. The planet is slightly gibbous. It is springtime and the retreating cap is fringed by an irregular dark collar. This is different and much darker in green light, indicating a markedly reddish colouring; the cap is somewhat enhanced. The red view shows what appears to be a yellow cloud over Argyre.*

Already the filter photographs of Mars taken in the 1920s by W. H. Wright and Dorothy Applegate at the Lick Observatory, in the U.S.A., revealed that the surface detail, and especially the maria, stood out dark and clear when a red or orange filter was used, but was washed out more or less completely through a blue or violet filter, while an ultra-violet one showed only the atmosphere. The diameter of Mars was substantially larger in ultra-violet than in infra-red exposures, and, since infra-red cuts through atmospheric haze, it was assumed that the difference in diameter corresponded to the depth of the scattering atmosphere of Mars, which was put at 60 miles. This is the so-called *Wright effect*, the reality of which has since been questioned.

At that time, however, the filter technique was still undeveloped, and in any case planetary photography is too crude a tool for judging minor gradations of shade or colour, which can be estimated with far greater accuracy by the human eye. Much better filters became available after World War II, and the visual use of tricolour separation sets seems to have been largely my own idea; in any event I specialised in this type of work.

The general effect of tricolour filters is well illustrated in the drawings prepared by the British amateur, A. W. Heath, using a 12-inch reflector, in the 1965 apparition (Plate VII). The extinction in green of the tip of Syrtis Major, which is intensely dark in red, is particularly striking and clearly shows the presence of a strong green or blue-green element in that part of the mare. In my own drawings of Mare Sirenum of 20 August 1956 the red filter revealed dark wedges which were almost invisible in green, while the dark collar round the south polar cap was very strong in green and pale in red, indicating a marked reddish coloration (Fig. 5), which comparison with a colour chart viewed through the same filters showed to be madder or maroon.

During the same opposition G. P. Kuiper observed Mars with colour filters, using a binocular attachment with the 82-inch McDonald reflector (Texas), which gave a linear magnification of 900 times. He states in conclusion that

Martian colourings, although subdued, are undoubtedly real.

There was a 'touch of moss-green' in the equatorial maria and of blue in the great wedge of Syrtis Major. Kuiper's estimate of the colour in the dark 'collar' round the south cap seems to be less red than mine, and he classes it as 'brown'. I suspect, however, that, being a painter in water-colours, I may be more discriminating in the estimation and naming of colours, so that there is no real difference here.

From visual observation without filters Clyde Tombaugh of Flagstaff (now Lowell) Observatory has distinguished three shades of green: a dirty brownish green in the Wedge of Casius (see map), a bright grass-green in the southern ('equatorial') maria and Mare Cimmerium in particular, and a strong bluish green – more blue than green – in Syrtis Major. His description of the 'collar' is 'a warm coffee'. All of this is in satisfactory agreement with the results detailed above.

During the favourable apparitions of 1956 and 1958 I found that the colour of the maria varied considerably, depending not only on the Martian season, but on the condition of the Martian atmosphere as well, at times the maria showing equally dark in red and green light. On the whole, however, they were distinctly greenish, although I have also seen a strong red, almost crimson, tone in Mare Chronium and some other subpolar regions.

The polar axis of Mars is tilted at 23°59′, which results in climatic zones very similar to ours and somewhat intensified seasons (Fig. 7). This similarity extends as we already know (p. 17) to the length of the day, which on Mars is equal to 24 h 37 min 23 s of our reckoning, and the seasonal resemblance is emphasised by the fact that, like the Earth, the planet passes through the perihelion in the northern winter (southern summer) half of the year. The effect of this situation, however, is much enhanced by the higher eccentricity of the orbit (p. 4), which brings Mars to within 128 200 000 miles of the Sun at perihelion and takes it away from it to 154 500 000 miles at aphelion. This entails a change in the allowance of sunlight from 0·52 to 0·36 of our solar constant, with a

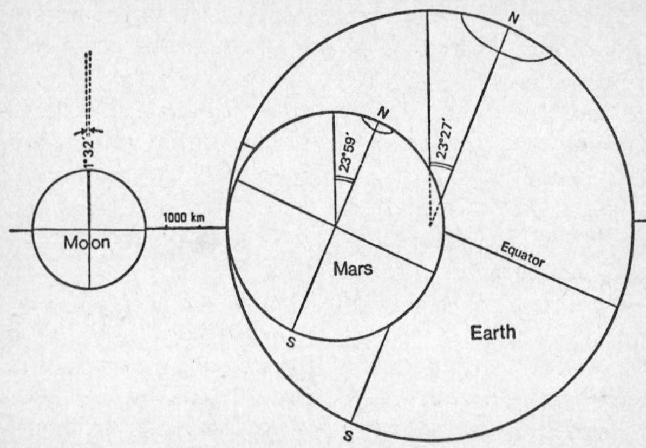

Fig. 7. *Moon, Mars and Earth compared. The horizontal line represents the orbital plane, and in the case of the Moon the plane of the ecliptic. Rotational axes and equators are also shown, as well as the angles between the normals to the orbital plane and to the equator (axial inclination or equatorial obliquity). The equatorial diameters are 3480, 6790 and 12 765 km respectively. North polar caps are indicated on Mars and the Earth.*

corresponding effect on the seasonal temperatures. Furthermore, since the year of Mars makes 687 of our days (roughly 668·6 Martian days), all seasons are lengthened in proportion, which has a cumulative thermal effect.

The perihelic approach to the Sun, during which Mars is speeded up in its orbit, falls in the southern spring and northern autumn, so that they are absolutely the shortest as well as relatively the warmest seasons of 146 Earth (or 142 Martian) days each. Contrariwise, the aphelic spring of the northern and autumn of the southern hemisphere are 199 Earth days (194 Martian days) long and relatively cold, while the southern summer of 160 days (156 Martian) is absolutely the warmest season, and the simultaneous northern winter milder than its southern counterpart of 182 terrestrial, or 177 Martian, days in length. The latter must be the bitterest time of the year anywhere on Mars, while the northern summer is comparatively cool.

All this is not without a bearing on the telescopic appearance of the planet, which closely follows the march of the seasons.

Mars is only 0·15 of the Earth in volume, which varies as the cube of the radius, but the superficial area of a sphere is proportional to the square of the radius, and is in the case of Mars equal to 55 million square miles, which, there being no seas, is almost exactly the same as the land area of the Earth (57·5 million square miles). Thus the scale of things is not inconsiderable by our standards.

On the average about 20% of these 55 million square miles is occupied by the maria and other dark features, 5% by the polar caps, while the 'deserts' engulf the rest. All these percentages, however, vary with the seasons. The dark markings shrink and grow pale with the coming of winter, as the polar cap creeps into lower latitudes, to withdraw again when the corresponding pole tilts towards the Sun.

The northern winter is the milder of the two, and the north cap at its maximum extent reaches down to 50° of latitude, which gives it a diameter of 3100 miles. By the autumn, however, this will have shrunk to about 200 miles; the cap will have grown dirty and barely visible, soon to be enveloped in mists and clouds, which cause it to project beyond the limb of the planet, especially when viewed or photographed in blue. Behind this veil the cap starts growing again, back to its icy splendour.

At the other pole the story is much the same, with time sequences reversed, but the winter is cold and bitter in the Martian antipodes and builds up the cap to a greater size. In 1954 it touched 44° of southern latitude, and in 1922 even 40°. Its average greatest spread is estimated at 3700 miles. This is not unlike the Earth, for the snows of our winters extend to comparable latitudes, but, despite the enfeebled sunshine, the rout of the Martian winter is much faster and more complete when the spring returns. This goes to show that the white covering must be very thin, most probably no more than a coat of hoar-frost, not exceeding a few inches in depth, except in the central parts. So the southern cap recedes rapidly in the warm perihelic Sun, becomes ragged round the edges and breaks up into isolated spots and patches. Its last remnant lingers in the so-called Mountains of Mitchell, which have never been seen as such, some 250 miles away from the true pole, and eventually disappears, though it may be re-formed at any time when white clouds overrun the pole.

The retreating 'snows' are fringed by a dark band of the so-called 'collar', already referred to on p. 25, and sometimes a thin blue line, as though of liquid water reflecting the sky, has been seen at their edge.

At the same time a 'wave of quickening', to use Lowell's graphic expression, progresses from the pole towards the equator. The dark areas deepen in shade, often altering in colour, and expand. It is an astronomical convention to use names from Classical Antiquity for planetary topographies, and this is how the map of Mars (see Plate IV) has become sprinkled, chiefly by Schiaparelli and Antoniadi, with *Phrygias*, *Syrtes* and similar inscriptions. One of these – Pandorae Fretum, in the southern hemisphere – refers to a region that, according to E. C. Slipher, extends at the height of summer to 1 000 000 sq. miles, but vanishes almost completely during the winter.

Other changes are less spectacular, but no less marked. The darkening advances at the rate of 10–30 miles per day through the maria and such minor features as *laci* (lakes), *luci* (groves), 'oases' and the controversial 'canals', as the long streaky markings, mainly between the maria, but sometimes entering and crossing them, have become known, up to and beyond the equator, invades the opposite hemisphere to a depth of 1000 miles or so, and then begins to retreat, as the same drama is re-enacted at the other pole.

E. M. Antoniadi was a highly skilled and diligent observer of the planets, having the excellent 33-inch refractor at Meudon, near Paris, at his disposal. During the latest perihelic oppositions of 1954 and 1956 the atmosphere of Mars was in a highly disturbed condition, great tracts of land remaining obscured for days and weeks on end by 'yellow clouds'. In 1924 and 1926, however, the Martian weather was much more favourable, which allowed the events on the ground to be followed in more detail. It is during these apparitions that Antoniadi observed some remarkable transformations in the colouring of the maria, bearing witness to considerable diversity in the nature of the terrain and so making any simple explanation the more difficult. The changes associated with the Martian spring were: (i) from grey or green to brown; (ii) from grey, green or blue to purple brown; (iii) from the same colours to maroon; (iv) from grey to lilac or a 'beautiful carmine hue'. Other areas remained unaffected, however, and stayed greenish or bluish, only darkening or fading. Antoniadi also notes brown, red and yellow colourings of short duration appearing in the maria without any recognisable cycle. In my own filter observations I have likewise noticed colour effects, which, though not detectable directly, must be interpreted as due to a carmine hue, 'beautiful' or not.

Such colour estimates are not easy, but Antoniadi was exceptionally well placed in this respect, and there can be no reasonable doubt that at least some colourings are affected by the seasons. While the intensification of some markings may be due simply to the removal of a light frosting or mist, forming during the winter months, the cycle of the observed changes appears to be associated with the release of moisture locked in the polar caps. Most of the maria are concentrated in a belt girding the planet at a mean latitude of 20° S, which may be plausibly attributed to the greater relative warmth of the southern summer, while the alternative preference for subpolar regions (Maria, Australe, Chronium and Acidalium) may be similarly related to the availability of water from the caps. All this points strongly to the organic nature of the colourings and their changes. In other words, the maria would be the vegetated areas of Mars and not simply dark plains, like the maria of the Moon, although until recently (see p. 54) it was generally believed that they were less elevated than the bright terrae, the difference of levels being of the order of 2000 ft.

The changes of colour are accompanied by variation in the proportion of polarised light, which increases with the intensity of shade, although the general shape of the curve remains unchanged for any given mare (Dollfus, 1961). Laboratory experiments appear to indicate that the polarisation is due to small objects, about 1/100 cm in diameter.

From the earlier polarimetric observations by B. Lyot and A. Dollfus it was inferred that the surfaces of the Moon, Mercury and Mars were covered with dust, and it has become a practice to regard the low-lying veils of a reddish and yellowish colour in the atmosphere of Mars as dust-storms. Yet Lyot found that these veils, referred to as 'yellow' or 'red' clouds, differed drastically in polarisation from the underlying 'deserts', and E. C. Slipher reported darkenings of the surface after it had been obscured by yellow clouds, which would be difficult to explain in terms of a dust-storm. Finally, the recent close-up photographs of the Moon have shown its surface to be substantially free from dust, which casts grave doubts on the original interpretation of the polarimetric results in the case of Mars as well.

In addition to the seasonal fluctuations in size, shade and colour, the dark features of Mars are subject to irregular changes, some transitory, others enduring. Thus during the opposition of 1956 Slipher observed the appearance in high latitudes of 'large shadowy dark areas which changed markedly from day to day', while on 8–13 September parts of the bright desert in the vicinity of Solis Lacus 'turned as dark as any markings on the disc'.

Permanent changes in the outlines of Solis Lacus and Mare Cimmerium have been recorded in the photographs taken at Flagstaff by P. Lowell and E. C. Slipher. The tip of the dark bay of Syrtis Major, which used to be pointed, has assumed a square-cut appearance in recent years. In the nearby region of Nepenthes-Thoth an area 'twice the size of Madagascar' has according to W. M. Sinton, changed its colouring from that of a desert to a mare since 1952. The peculiar thing about these 'permanent' changes is that they seem invariably to result in the extension of the dark areas at the expense of the bright. In any event the 'geology' or topography of the surface could not possibly vary in this way in so short a time.

4. What About the Canals?

The Mercator map of Mars produced by the International Astronomical Union (see Plate IV) shows no 'canals', although some N.A.S.A. maps do, and it will have been noticed that only a passing reference has been made to them in the foregoing chapter.

If I am allowed a personal touch, let me confess that I have never seen anything like a 'canal' in the meaning of a thin spidery line running along a great circle of the Martian globe (for this term is applied also to a few broad streaks which are readily visible).

We all know that 'seeing is believing', optical illusions and hallucinations apart. Thus it is perhaps not unnatural that astronomers often tend to approach the problem from the standpoint of personal experience, for scientists are not a special brand of mankind; they are not free from ordinary human failings, however honestly they may strive at objectivity and circumspection. Indeed, these very virtues may sour into the curmudgeonly rejection of anything likely to disturb the slumbrous rhythms of astronomical routine.

I have never looked at Mars through anything bigger than an 8-inch refractor or a 12-inch reflector (my own), and, although 'canals' have been recorded with much smaller instruments, I am prepared to listen to those better equipped or more favourably sited under clearer skies or on high mountain tops.

A few 'canals' appear in the drawings made by J. H. Mädler in Germany with the aid of a 3½-inch refractor in 1840, but they were properly discovered by G. V. Schiaparelli of Milan during the perihelic opposition of 1877, which also brought the detection of the midget moons of Mars by Asaph

Hall at the U.S. Naval Observatory in Washington. It is something of an irritating cliché that Schiaparelli in announcing his discovery called these features *canali* in Italian, the correct English for which is 'channels', not 'canals'. Yet 'canals' they were to become and have been known as such ever since.

The puzzling geometrical pattern of these lines soon engendered the idea that they represented a stupendous irrigation system, constructed by superlative engineers, to husband the meagre supply of water locked in the polar snows on a 'world athirst', as Mars was described with no lack of imagination by the wealthy American enthusiast Percival Lowell.

Lowell founded in 1895 at Flagstaff, under the blue skies of Arizona, a well equipped observatory for the express purpose of studying Mars and other planets – but Mars mainly. As I write these words he seems to be looking at me with a pair of honest blue eyes out of a handsome earnest face above a tie somewhat out of touch with the collar in one of those faded brownish photographs with yellow edges. His initials P.L. are immortalized in ♇, the symbol of Pluto, which he had tracked down mathematically from the orbital perturbations of Uranus and Neptune, but he died in 1916, 14 years before the new planet was actually found in an earlier sky photograph by Clyde Tombaugh at Flagstaff.

The observatory is now called Lowell. The U.S. Air Force is mapping the Moon and watching the planets. The times have changed, but Mars has not, and the problem of its 'canals' is still with us, for all the bottles of ink expended on it over decades of controversy.

Schiaparelli was able to confirm Lowell's findings during the oppositions of 1879 and 1881, and in 1886 the first independent observations of the 'canals' were made by Perotin and Thollon at Nice. Flagstaff became famous over the years, and much valuable work continues to be done there.

As represented in a typical Flagstaff drawing by Lowell *et al.* from the early days (see Plate VIII), the 'canals' are thin lines of uniform width, running quite straight along great circles for hundreds and thousands of miles, from mare to mare, which they enter and leave by triangular 'carets', and the poles. They meet and intersect, up to 14 at a time, in minor dark spots, christened 'lakes', 'groves' or 'oases' and measuring 75–100 miles across. To be visible at all, the 'canals' themselves must be at least 15 miles wide.

They were said to participate in the general behaviour of the dark markings, fading with the approach of winter, to darken again as the 'wave of quickening' swept over or down them from the poles in spring. They were, however, also subject to seemingly capricious changes, unrelated to seasons, disappearing for several years at a time and altering in width or intensity. The doubling, or *gemination*, of the 'canals' was first noticed by Schiaparelli and repeatedly observed at Flagstaff, where some canals have even been seen to triplicate. The component parts of a geminating 'canal' have been reported to pursue a geometrically accurate parallel course, from 75 to 400 miles apart. Such an appearance could be due to optical factors – say, double refraction. But it was pointed out by the Flagstaff astronomers that this would affect all the 'canals' simultaneously, and not only some of them in any given region, as observed. Moreover, the effect was strictly limited to the equatorial part of the planet. This appears to rule out the optical explanation.

By the time Lowell's *Mars as the Abode of Life* (p. 23) went to press in 1906, 437 'canals', 51 of which double, and nearly 200 'oases' had been mapped at Flagstaff.

A world-wide system of so regular a nature seemed far too purposeful to be attributed to chance natural causes, and Lowell was deeply convinced that it was the work of a highly intelligent race of beings, not only far in advance of us in technological development, but also able to sink their differences in peaceful co-operation for the common weal. A conception so sensational, if perfectly logical granting the correctness of the observational data, had a better reception with the popular Press than with the scientific circles.

Various objections were raised. The 'canals' were far too wide to be artificial waterways. For one thing, there was not enough water on Mars to fill them. To this, the obvious answer was that what we saw was the vegetation springing along their banks rather than the canals themselves. Then, since the poles are depressed below the equator owing to the strong oblateness of Mars (p. 17), the water from the polar caps would be required to flow uphill. This could, no doubt, be partly dealt with by locks or similar installations. But any open water would be subject to rapid evaporation under Martian conditions, as we shall see in another chapter, so that distributing the precious liquid by means of canals would be a very uneconomical way of irrigating the desert planet. It was more logical to assume that water from the poles was conveyed by pipelines equipped with pumps. In other words, there would be no canals at all, just strips of vegetation.

So long, however, as the reality of Lowell's picture of Mars was admitted, there was no easy escape from his conclusions. To overthrow these, it was necessary to deny the accuracy of the observations on which they were based, and this is what most of his opponents did.

Some astronomers continued to show the 'canals' as depicted by Schiaparelli, if not perhaps by Lowell. Others professed to be unable to see them at all, or at most only as hazy streaks.

Among the latter was the famous American observer of proven acuity of vision, E. E. Barnard, who used the 40-inch telescope at Lick, the largest refractor in the world, as well as the 60-inch reflector at Mount Wilson. By means of these he descried much intricate detail in the maria, and left us the following description of Mars:

> '. . . A globe whose entire surface had been tinted a slight pink colour, on which the dark details had been painted with a greyish-coloured paint, supplied with a very poor brush, producing a shredded or streaky and wispy effect in the darker regions' (quoted after *Astronomy* by Russell, Dugan and Stewart).

But Barnard could not see any 'canals'. It was claimed in explanation that the definition of such large instruments was inferior to that of smaller telescopes, because the larger aperture gathered light from a wider area of the sky, which was less likely to be undisturbed, and so long as the refractive imperfections of the air exceeded in scale the telescope's field of view the image might be shifted as a whole, but was not disrupted. The argument is sound in theory and practice, and it is often necessary to stop a large instrument down to a lower aperture in observing the planets, to reduce 'boiling' (p. 7) But, as we shall see presently, this is not the whole story.

Antoniadi and most British and Continental observers, while not denying the reality of linear markings on Mars, saw them as neither so straight nor uniform in width, as did Lowell, and maintained that in best seeing the 'canals' tended to break up into strings of disjointed detail. It is, indeed, an established fact that the human eye, when faced with faint objects near the threshold of resolution will often construe them into rectilinear patterns.

To test this idea, the Astronomer Royal Maunder prepared a large drawing of Mars, in which he replaced Schiaparelli's 'canals' by wiggly lines, indistinct shadings and spots. He then invited the pupils of the Greenwich Hospital School, ignorant of the true appearance of Mars, to copy his drawing from a distance, without allowing them to inspect it at close quarters. Several of the copies showed unmistakable 'canals'.

This is substantially the position occupied today by Audoin Dollfus, a keen-sighted and diligent student of the planets, having at his disposal the 24-inch refractor at Pic-du-Midi, the highest permanent astronomical observatory in the world (9351 ft above sea-level). He, too, finds that the 'canals' break up into spots as the seeing improves. Nevertheless, he says that these concatenations of fine detail 'constitute a very peculiar and specific characteristic of the Martian topography'.

Here it may be observed that the discontinuity of the 'canals' in which their general linear character and geometric distribution are preserved does not invalidate Lowell's conception, for if the water supply is carried by pipelines it would be unreasonable to expect it to be loosed on the way regardless of the terrain, which might be locally unsuitable for cultivation. Thus we would have only dark patches of vegetation here and there along a pipeline, invisible as such.

There is, however, independent evidence that the 'canals', whatever they may be, contain moisture.

Thus during the 1954 opposition a W-shaped cloud was observed and photographed to form every Martian afternoon in rough coincidence in its bright 'blobs' with the 'oases', Arsia, Silva, Ascraeus Lacus, Tithonius Lacus and Hebes, and in two of its strokes with the 'canals', Ulysses and Fortunae.

Furthermore, the present director of the Lowell Observatory, John S. Hall, maintains that, contrary to Dollfus's contention, the fine 'canals' are visible *only* at best seeing. E. C. Slipher's book, *Mars – The Photographic Story* (1962), contains an interesting juxtaposition of composite photographs of Mars, where several negatives taken at short intervals by himself and H. L. Giglas have been printed on top of one another, to defeat atmospheric unsteadiness and capture the fleeting moments of perfect seeing, and placed side by side with the observational drawings of Mars prepared by Patricia Bridges with the planet in the same orientation and at the same season. The latter display a number of specially emphasised 'canals', including Nepenthes-Thoth, which may also be seen in A. W. Heath's drawings in Plate VII. The half-tone reproduction of photographs inevitably involves some loss of finer detail, but the original prints are in my possession, and I can testify that most of the 'canals' appearing in the drawings can be traced in these as well.

Another kind of test is provided by the cases of Edison Pettit and R. S. Richardson, neither of whom had believed in the existence of Lowellian 'canals', and then suddenly saw them quite plainly: Pettit in 1939 and Richardson in 1956.

I will preface this story with a short extract from Richardson's *Man and the Planets*, published in 1954, where he records how in October 1941, when the seeing was very good, he searched for the 'canals' with the 100-inch Hooker Telescope and a power of 1000 diameters. 'But,' says he, 'like the old lady who could not see the ice cubes, I was still utterly unable to see any canals.' He fared no better with the 6-inch,

12-inch and 60-inch telescopes, so that this was not a matter of aperture.

On 3 June 1956, however, using the self-same 60-inch reflector as had been employed by Barnard, Richardson had the following experience:

> 'This morning the disc had a peculiar aspect which I had not noticed before. There were innumerable irregular blue lines extending across the bright red regions like veins through some mineral. Several minutes passed before it occurred to me that these markings must be canals. I was taken completely by surprise, as I had not thought of seeing canals at a distance of 75 million miles. . . . These lines appeared distinctly blue, the same colour as the maria. In fact, they appeared to be narrow extensions of the maria into the deserts.'

Pettit saw the 'canals' olive-green, while most other observers did not put any colour to them at all, but these are minor points. More important is the observation that the 'canals' were irregular, like the veins in a mineral, rather than a geometrical system. From this Richardson concludes that they must be natural 'geological', or rather *areological*, features.

Clyde Tombaugh has suggested that the 'oases' may mark the points of impact where asteroids hit the planet, fracturing the surface in spoke-like patterns around them. This is not impossible, but there is no evidence of such scars on the Earth, and the detailed study of the Moon gives little support to the hypothesis that its craters have been formed by impact. It has also been proposed that Mars, having been formed cold (p. 13), later expanded through radioactive heating and its surface cracked open in a world-wide pattern, to keep pace with the increased volume of the interior.

It is easy but not very profitable to speculate about an unverifiable past. On the other hand, there exists on the Moon a globe-wide system of tectonic lineaments, known as the grids, which generally tend to follow the great circles and may extend for hundreds and even thousands of miles in nearly-straight lines, now manifesting themselves as faults, rilles (grooves) or crater-chains, now as ridges and boundaries of ring complexes and maria (J. E. Spurr, V. A. Firsoff, G. Fielder, K. v. Bülow, and others). A similar fracture system is traceable on the Earth as well (Vening Meinesz, W. H. Hobbs). It is possible but conjectural that these grids are due to the tidal interaction between the Earth and the Moon. Since, however, Mars occupies in many respects a position intermediate between these two bodies, it seems logical to infer a similar tectonic grid or grids on Mars. With erosion more intense than on the Moon, coupled with a general dryness of the surface, the fractures would be expected to gape wider in time, acting as natural recipients of moisture and possibly giving a foothold to vegetation. This is, indeed, what I suggested, without any great claim to originality, in 1963.

Thus there could be 'canals' without Martians, or, conversely, Martians without 'canals'. In fact, if the hypothetical inhabitants of the planet were as clever as they were 'painted', they could probably obtain water much more easily from underground by sinking artesian wells. The 'canals', natural or otherwise, could then be not so much an irrigation as a communication network. E. C. Slipher, who was observing Mars from South Africa in 1954, reported that the 'canals' sometimes crossed each other, which is rather puzzling, although the effect could be due to one fracture being trenched deeper than the other. Gemination is not so easy to explain.

Great hopes were entertained that the American Mars probe *Mariner 4* would solve the problem of the existence and nature of these 'canals'. Yet, successful as it was in photographing the planet, it brought us only views of the winter hemisphere of Mars, and all dark features, including the 'canals', tend to fade out in winter. Be this as it may, the photographs do not seem to show any 'canals'.

5. Mariner 4

The Moon probes have staged up a confrontation between the results obtained by relatively distant telescopic observation and the hypotheses deduced from these, on the one hand, and actuality, on the other. The confrontation continues, but it has already shown that volcanic processes have intervened importantly in the shaping of the lunar surface, while the impact conception according to which the Moon has been merely a passive recipient of cosmic bombardment, mainly in a remote past, has not fared at all well. This is not without a bearing on Mars, which resembles the Moon much more than it does the Earth.

The Moon, however, is only some 240 000 miles away and, being a satellite of the Earth, is not in the same category as Mars, an independent planet with an atmosphere and indications of life on its surface. This is why *Mariner 4* is a great landmark in the exploration of the Solar System.

The probe left the launching pad at Cape Kennedy atop an *Atlas-Agena* rocket, used to put it into orbit, on 28 November 1964, to pass at about 6000 miles from Mars on 14 July of the following year. At that time the planet was some 135 million miles away, and so not well placed for radio communication. The situation would be vastly more favourable if the fly-by occurred at opposition, as radio signals suffer attenuation in proportion to the square of the distance.

Apart from recording cosmic radiation, micrometeorites and magnetic fields, the probe was designed for a close-range photographic reconnaissance, for which purpose it carried a camera and a small reflecting telescope of Cassegrainian design with a beryllium primary mirror 1·62 inches in diameter and 12 inches in focal length. The shutter of the camera

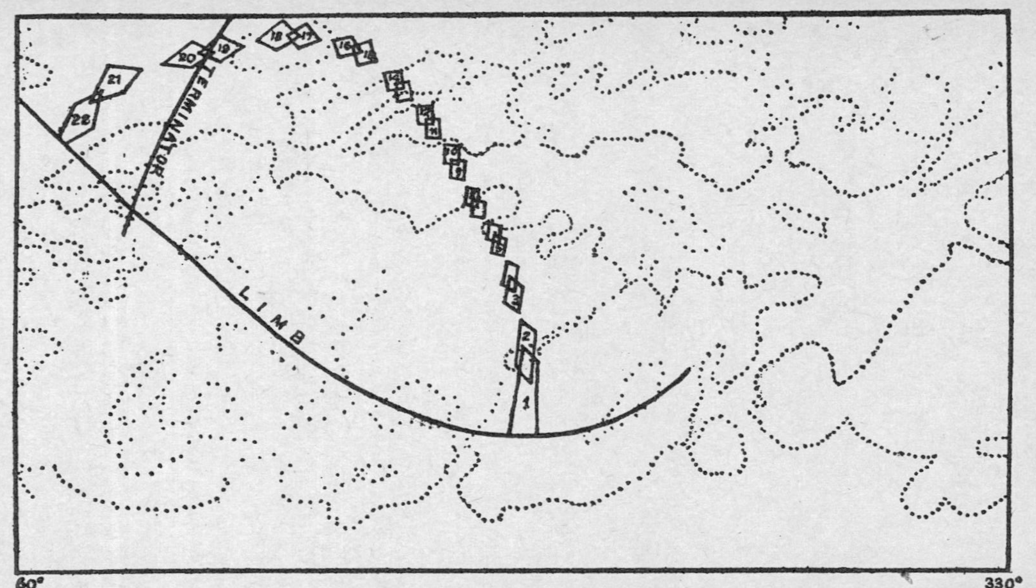

Fig. 8. *Distribution of the* Mariner 4 *frames on the surface of Mars. Compare with Plate IV (map).*

consisted of a rotary disc, mounting four filters in alternate pairs of blue-green and red-orange, to study colour effects. Twenty-two pictures (Fig. 8) were taken all in all, in pairs with partial overlap at 24-second intervals, with an exposure time of one-fifth of a second.

It was winter time in the southern hemisphere of Mars, and the ground covered by the photographs extends from the desert region of Amazonis near Trivium Charontis, which lies at about 20° of northern latitude, through Zephyria across the equator into the southern tropical maria, Sirenum and Cimmerium, which, though, were only skimmed at the edges, most of the coverage falling within the bright strip of Atlantis between them. Thence the 'eye' of the probe swept east into another 'desert' of Phaethontis and the subpolar dark area of Aonius Sinus near the morning terminator, passing on into the night (see Plate XVI).

The great distance from the Earth did not permit direct transmission by television, as in the case of the lunar *Rangers*, and a form of facsimile telegraphy was used instead. The pictures were formed on a videcon tube and scanned by 200 lines, every line containing 200 elements, graded into 64 shades, from full white to full black. The result was recorded on magnetic tape as 'bits', each bit being a binary digit that could be converted to a shade and a position and the whole picture reconstructed in this way bit by bit. There were about 250 000 bits per photograph, and these could be 'played back' to Earth at relative leisure of 8⅓ bits per second, after the *Mariner* swung round Mars into the Earthward leg of its orbit. In fact, it took nearly ten days to transmit the information corresponding to the 22 photographs, taken at slant ranges (the camera was not looking straight down at Mars) from about 10 500 to 7600 miles.

The last seven photographs show little or no detail, be this due to a low level of illumination, fogging of the camera or mistiness near the terminator, which was crossed in frame 19. The remaining frames, however, have brought some remarkable, although tantalisingly inconclusive information, which is still being 'evaluated' at the time of writing.

The Russian Mars probe *Zond 2*, launched two days after the *Mariner*, failed before reaching the planet.

Exterior planets are generally easier to observe than the interior ones, as they can be seen at opposition fully lit when closest to the Earth (p. 7). On the other hand, they can never be seen at a small phase, when the unevennesses of the ground cast long shadows near the terminator, whereas the full-phase view is essentially devoid of shadow, the Sun being directly behind the observer. For this reason the surface relief of Mars does not show to advantage in a telescope. The arc of shadow never encroaches on the visible hemisphere by more than 47°, which makes only a thin slice through spherical foreshortening. Moreover, Mars at such times is far away and correspondingly difficult to study.

Thus not much was known about the sculpture of the Martian ground from telescopic enquiry. It was suspected that the maria may be lying some 2000–4000 feet below the mean level of the bright terrae, mainly, it seems, by analogy with the Moon rather than on any observational evidence. The existence of gradual heights was inferred from the white patches left behind in the same places by the receding snow-caps, and we will recall the Mountains of Mitchell (p. 30). In the morning some points and areas near the terminator tended to whiten, which was interpreted as a deposition of hoar-frost in the cold of the night on hilltops or plateaux rising above the surroundings.

On the whole, however, Mars was thought to be an unsculptured world, without high mountain ranges or sharp differences in altitude, and the variation in brightness with phase, to which invisible shadows contribute, appeared to bear out this view.

True, E. J. Öpik expected craters to have been formed on Mars by meteorites hitting the surface, which was in line with Tombaugh's ideas about the origin of the 'oases', while in my book *Exploring the Planets*, published in England in 1964, I forecast fault-bounded 'block mountains' and low volcanic calderas from general geological considerations. None of these, however, had ever been seen.

It, therefore, came as a shock of surprise to most astronomers when the *Mariner* revealed Mars to be a world resembling the Moon rather than the Earth in surface structure, with a seeming abundance of vertical features and even high mountains, although it must be borne in mind that the 22 photographs obtained by the probe account for only 1% of the surface, and some of them, as we have just noted, show nothing at all, which leaves plenty of room for further surprises.

However, the best plan is to examine the pictures themselves.

Frame 1, taken at a mean range of 10 500 miles, is centred at 35° of northern latitude in the region of Propontus II and Phlegra (see map, Plate IV). It gives a slanting view of the Martian surface towards the limb, about 400 miles away from the bottom of the frame. The picture has been printed in three degrees of contrast, and the lower contrast prints show a thin cloud above the horizon, officially described as a 'defect'. The cloud is very high up over the surface and could be one of the *blue clouds* to be discussed later on (p. 69). Most of the bright detail in the high-contrast print closely resembles cloudscapes in the satellite photographs of the Earth and is most probably atmospheric. The ground features include curious thin streaks with a general trend from the south-west to the north-east, which we shall encounter in the further frames as well. But there are also indistinct dark linear markings on a much larger scale and differently orientated, which could possibly be interpreted as 'canals'. The next frame again shows bright cloud-like detail, but also a suspicion of shallow ringed depressions, as well as the mysterious streakiness, which appears to have two main components approximately at right angles to each other and inclined at 30–60° to the scanning grid, so that it cannot be due to it.

In the following pictures up to No. 6 the detail is meagre, as the Sun is about 20° from the zenith and the shadows are short. Both crateriform and more extended depressions (frame 3), as well as the streaky grids can be glimpsed. All of them stand out with increased clearness in frame 7, taken at a mean range of 8400 miles with the Sun about 60° above the horizon and so longer shadows, through a blue-green filter. The walls of the ringed depressions appear to be low and gentle and the depressions rather shallow for the most part, some having been eroded almost out of existence. There are some strong highlights in the northern part of the picture. The locality is Zephyria, just south of Mare Sirenum. The next frame, taken through a red-orange filter in partial overlap with frame 7, penetrates the mare. The relative extinction of highlights suggests that these are largely atmospheric. The salient features of the view are three craters, 25–40 miles in diameter. They are flat-bottomed, have hardly any outside walls and bear a general resemblance to Ngorongoro in the Crater Highlands of Tanzania or the maur craters of Uganda. A chain of craters up to a few miles across extends in an arc westwards from the large ring in the north-east. The largest ring below the centre is distinctly hexagonal with a suggestion of terracing to the west of it.

In frame 9 we are in Mare Sirenum. The Sun stands 38° from the zenith and throws the ground sculpture into clear relief. The whole area of 160 × 170 miles is pock-marked with roundish depressions, one of which has a large central peak and another scalloped edges elevated above the surroundings, but most of the craters seem to be of the 'ghost' type, well known from the lunar maria. At least one of the streaks appears to break up into a chain of small craters, also a familiar lunar feature, but the nature of the abundant minor detail is obscure and lost in the scanning grid.

Frames 10 and 11 fall within the bright strip of Atlantis that divides Mare Sirenum from Mare Cimmerium. This seems to be mountainous country, in which large rings up to 75 miles in diameter abound. In No. 10 a light atmospheric veil is suspected, but No. 11 is the showpiece of the series (Plate XV) and deserves closer attention.

The range is 7800 miles, the altitude of the Sun 43°. The blue-green filter having been used, the emphasis in the dark features is on the red, and the highlights are shown prominently.

The first impression is lunar, and yet to a practised eye it is not the Moon. It is not just the lack of the crispness and high contrast of lunar photographs arising from the transmission technique. The outlines are genuinely softer. The shadows are not so deep owing to atmospheric scattering. The erosive forces are more effective. It is a place where winds blow and there is at least some water – a bigger, more mysterious world.

Most of the view is filled by a huge ring structure, some 75 miles in diameter. The eastern and southern walls have largely succumbed to erosion and the floor of the depression seems to dip towards the north-west, where the walls reach considerable height. Several small craters can be made out along the rim, and one of these marks the summit of a conical peak – an unmistakable volcano. In the north a mountain complex, flanked by large calderas, overtops the outer *glacis* (slopes) of the 75-mile crater, and flows of lava descend in tiers to its floor, from which a round dome bulges up. There may be a craterlet at the top of the dome, or else this may be a grid mark, but there is no mistaking the two craters on the lip of the north wall, whence the lava seems to have come.

In the north-east, partly encroaching on the big ring, a caldera about 18 miles across is festooned with rim craters, marking the ring fracture along which volcanic activity broke out after the collapse of the dome that must have originally occupied the place of the central depression. Such features are well known on both the Earth and the Moon and bear a clear mark of igneous activity. A winding *rille* (a lunar term and the German for 'groove') can be traced southwards down from the glacis of the caldera and may be a watercourse.

We are in the winter hemisphere of Mars, and the highlights in the west and north may be due to hoar-frost, which is emphasised by a green filter, and possibly light clouds over the high peaks.

There is much small detail within the 75-mile ring, but it does not show clearly enough to attempt positive identification. It is possible that by varying the contrast of the print (which can be done almost *ad lib.* by television technique) something more can be learned about it, but if this has been done the results have not been published to date, nor are any such prints available to me.

There are two deeply-trenched craters within the big enclosure. Both of these are deeper on the north-east than on the south-west side, thus confirming the general lie of the land, and the larger of the two has a rectangular outline in the west, which again indicates deforming forces acting from this side. Indeed, the ground is fractured and faulted here and chains of craterlets are suspected. The sides of the rectangle are in sympathy with the general 'streakiness' of the surface, which shows that this must be tectonic in origin. In the east the glacis of the main ring are cut by fault-lines. A nearly straight line can be followed right across the photograph, trending from WSW to ENE. It could be a crater-chain, or a source of vapour, and, when fringed with vegetation in summer, simulate a 'canal'.

Frame 12 from Mare Cimmerium gives an impression of gently undulating, but not strongly accidented terrain, but mountains reappear in frame 13.

Although the Sun stands only 33° above the horizon, the shadows are very pale, bearing witness to atmospheric haziness. Eroded rings show faintly in the north among a slanting checker-board of squares and oblongs. In the east large rings are marked out dimly by hoar-frost or cloud along the rims.

The main feature, however, is a linear scarp, running south-west to north-east at about 30° to the horizontal component of the scanning grid, with high ground to the south of it, where mountains rising up to 13 000 ft have been identified. So much for Mars being a world of plains and flat horizons!

The photographs numbered 14 and 15 refer to the 'desert' of Phaetontis. Ring complexes are picked out by highlights, especially as the filter is blue-green, and are most probably rim hoar-frost, as has, in fact, been suggested by the scientists in charge of the *Mariner* project. It is, however, a matter for surprise that they have failed to comment on the virtual absence of shadows within the rings, although the Sun is only 30° and 24° respectively above the horizon, and such shadows show clearly in the earlier frames, obtained in less favourable illumination. The only possible explanation is a further thickening of the ground haze or mist, already noticeable in frame 13, with the approach to the morning terminator. This consistent trend reaches its logical conclusion in the next frame, where all ground detail is more or less wholly obliterated, and, as already noted on p. 43, the further frames show nothing at all.

Local obscurations are familiar to observers of Mars, where surface detail may become invisible for days and even weeks on end owing to atmospheric veiling. Thus Ben Burrell, Deputy Director of the Mars Section of the B.A.A., also thought that the loss in contrast was atmospheric, without going into the matter in the way given above, and added that this region had been 'notorious for lack of detail in recent months' (i.e. before 14 July 1965).

Passing reference has already been made on p. 25 to the reflection curve of Mars. The drop in brightness, or increase in magnitude, is strongly marked on either side of the full in the case of the Moon, which at *dichotomy* (half-phase) gives only 8% of the light of the full disc. This is due to the shadows cast by its surface relief and the porosity of the ground. Mercury behaves similarly, but Mars reflects light very nearly as a smooth sphere, which is difficult to understand if its surface bears such a strong resemblance to the lunar. One reason may be that the Martian rocks are not so porous as those of the Moon or Mercury, but the situation can be readily accounted for by a ground mist developing at the evening and persisting after the night at the morning terminator, so that at the limb the reflection comes not from the ground, but from the atmosphere.

The official boffins, on the other hand, suggested that the drop in contrast was due either to the fogging of the videcon surface of the camera (why?) or the surface of Mars being much darker near the terminator than anticipated. The latter could happen if it were highly porous, but this explanation as a whole is too artificial and does not fit the sequence of events as described here in the review of the *Mariner* photographs frame by frame. It will be recalled that the decrease in contrast was gradual and affected the shadows on low ground first and only then the highlights pertaining to heights or the overlying atmosphere.

It is always a temptation to bend observational evidence to theoretical expectations, and in this case the trend of official thought seems to have been fathered by the idea that the atmosphere of Mars is too thin and holds too little water vapour for thick mist to develop. Yet a 'glow of vaporisation' near the morning terminator has been recorded by reliable observers, using first-class equipment; condensation of carbon dioxide, which is known to be present on Mars, could occur at low temperatures, and, finally, morning and evening breezes could raise fine dust veiling the surface.

The point is that, in addition to its photographic mission, *Mariner 4* has carried out an 'occultation experiment'. *Occultation* is an astronomical term used to describe the covering of a more distant celestial body by a closer one, and here the 'celestial body' was the *Mariner* itself. As it passed behind the planet, the radio signals emitted by it had to traverse the Martian atmosphere and were subject to refraction and attenuation on the way. By measuring these it is theoretically possible to determine the *scale height*, or the rate of decrease in density with altitude, the total mass and the ground pressure of the interposed atmosphere.

The subject properly belongs in another chapter. However, the earlier determinations, based on atmospheric scattering and polarisation of sunlight at the planet's limb, gave a ground barometric pressure of some 90 mb, or about 1/11 of our sea-level pressure, which corresponds to an atmospheric mass between a fifth and a fourth of the terrestrial. On the other hand, the figure obtained from the *Mariner* 'occultation experiment' is only between 0·01 and 0·05. But before accepting this result as superseding the earlier determinations several points must be considered. Thus, firstly, it is based on a comparatively weak signal with Mars 135 million miles away, whereas the scattering and polarisation measurements involve the full sunlight at opposition distances. Then, the refraction of radio waves depends on the electron density and not on the atmospheric mass as such, and the relationship between the two is a matter of theoretical and unverifiable assumptions. Finally, *Mariner 4* was also equipped with a solid-state charged-particle telescope, capable of detecting electrons with energies greater than 40 KeV (kilo-electron-volts) and protons of energies exceeding 1 MeV (mega = million eV). Yet no evidence of particle radiation of this nature has been found at 6000 miles from Mars, although such particles were registered on the outward flight up to 103 300 miles from the Earth. From this it is inferred that the magnetic moment of Mars is less than 0·001 of the terrestrial (J. J. O'Gallacher and J. A. Simpson). This again involves certain assumptions, but the lack of an appreciable magnetic field should affect the ionospheric structure of the Martian atmosphere and may be negatively reflected in its electron density, used in calculating the atmospheric mass.

To sum up, the 'occultation experiment' cannot be regarded as conclusive.

As for calculating the 'ages' of Martian craters from the assumed rate of meteoritic infall, it would have to be proved first that the craters of the Moon were so formed, and the evidence is overwhelmingly against this view. One point alone is quite sufficient to disprove it: wherever two lunar craters overlap, the overlapping crater is almost without exception smaller than the overlapped one. This would require each successive meteorite hitting a given area to have been of smaller mass and/or moving at a lower velocity, which is an impossible assumption. On the other hand, it is perfectly reasonable to expect internal forces responsible for crater formation to decline with time.

6. The 'Geology' of Mars

A planet is an organic whole, the outcome of thousands of millions of years of development, in which the contending forces have achieved a state of fine balance. Such a whole is much more than the sum of its parts, wherein lies the difficulty of reaching correct conclusions by the process of analysis. Theoretical models, in which an artificial whole is created by considering simplified factors in isolation, may be instructive in some respects, but are as often highly misleading, and we must beware of substituting them for the real thing and forgetting the facts of observation.

The same analytical difficulty crops up in any orderly classification, even such as breaking up a subject into chapters, because in an organic whole everything is interlinked. The orbit and axial spin affect the climate, which is complexly involved in atmospheric structure, and this in turn has geological implications. Gravity enters into it all at many points, and if there is a biosphere it, too, may affect the atmospheric composition (free oxygen, carbon dioxide, etc.), as well as the structure of the surface, to take only mineral coal and limestone as examples of biological rocks.

It is, nevertheless, possible to deduce quite a lot about a planet by recourse to simple principles in the light of relevant experience acquired elsewhere, and this applies in particular to areology ('geology' of Mars) where we can draw on the knowledge derived from the study of the Earth and the Moon.

The subject has, in fact, already been broached in Chapter 2 and touched upon at various points in Chapter 5. Thus we have seen that the overall composition of Mars could not differ greatly from that of the Earth, but with the balance tilted in favour of lighter constituents, the sialic layer being much thicker on Mars and probably completely enveloping the mantle to some depth. This makes the occurrence of basic rocks on the surface highly unlikely.

Now, our oceanic basins represent crustal depressions which are isostatically balanced against the upraised continental masses because they consist of heavier basic rocks, and these are probably buried about 100 or more miles below the Martian surface. It is therefore clear that Mars can have no analogues of our oceanic basins, perhaps not even of the lunar maria.

This is not the same as saying that there are no depressions on Mars. Differences of density exist within our sial and may be expected to be the more marked in the much thicker crust of Mars. The fact that there are mountains rising to 13 000 feet above the surrounding country, and perhaps even higher ones, clearly points to considerable isostatic inequalities. It must not, however, be forgotten that, as explained on p. 21, Martian surface rocks, even if mineralogically identical with ours – and there is room for inclusion of lighter components – will be bubbly or lightly compacted and correspondingly less dense, while everything weighs 2½ times less on Mars in any case. This makes isostatic compensation so much easier.

In these circumstances the most likely mechanism by which depressions could arise would be the densification of the light formations by degassing and consolidation through the action of internal heat, permeation by volcanic exudations or by water. If this happened they would not only shrink in volume, but be pulled down by their own weight. It may be that the lunar maria owe their origin to the same process (Firsoff, 1961).

It will, however, be observed that, whereas on the Moon the maria tend towards subcircular or polygonal outlines, as is natural for subsidences, the Martian maria are entirely different in shape, and it is quite possible that the only thing the two types of mare have in common is a dark colour. Indeed, the *Mariner* photographs do not appear to show any marked difference in altitude between the dark marial and the bright 'desert' terrain, though admittedly the bright strip of

Atlantis between Mare Sirenum and Mare Cimmerium is mountainous land. We may also recall that one and the same area may appear as a 'mare' or a 'desert' depending on the Martian season, and even change from the one into the other more or less permanently in a couple of years or so. Nothing of the kind happens in the lunar maria, nor for that matter in our oceanic basins. It has even been suggested, notably by Sagan *et al.*, that the maria of Mars may be elevated above the bright terrae, and this allows them to escape a dust coating, to which the colouring of the latter is due. We have just seen, however, that this does not apply to the mountainous Altantis, and it is difficult to see how such an hypothesis is supposed to account for the seasonal colour changes. The experience with the Moon indicates that the ideas of dust on other planets, so dear to many astronomical hearts, must be taken with a generous pinch of salt (p. 32).

The studies of radar reflections from Mars made with the 1000-foot fixed radio telescope at Arecibo, in Porto Rico, show considerable variations in the structure of the surface. According to R. M. Goldstein (1965), some areas appear to be quite smooth, but rough regions with both strong and low radar reflectivity in the wavelengths of about 12 cm have also been found. Two of the high-reflectivity areas are about Trivium Charontis and at the tip of Syrtis Major, both of them predominantly 'desert', but, *The Sky and Telescope* comments, what the two have in common is marked recent changes in appearance. Characteristically, however, Mars gives radar echoes different from those of the Moon and Mercury, in the case of which a similarity of surface structure is inferred, and appears to be much smoother. Presumably the porous 'overcooked-toffee' surface texture, typical of many lunar regions, is rare or absent on Mars.

Be this as it may, its surface is continental in the geological meaning of the term, resembling the unfolded sialic regions of the Earth known as *shields*. This is one more reason for the absence of seas, either at present or in the past, other reasons having been considered in Chapter 2 (p. 21 ff.). Such arguments, however, do not preclude the existence of lakes, although their survival is made very problematic by the high rates of evaporation due to low barometric pressures and the relatively high summer temperatures of the ground (see Chapter 8).

On the other hand, *underground* lakes and rivers are not uncommon in the arid regions of the Earth. On Mars, as we have seen (p. 21), the existence of a permafrost seal, combined with the low density of the rocks, which could even be, at least locally, specifically lighter than water, and in any event would fracture readily, should favour such developments. Indeed, most of the hydrosphere may be permanently locked in the interior, only small amounts of liquid or 'precipitable' water being released to the surface during volcanic and tectonic disturbances, in hot springs or geysers, to be rapidly evaporated or reabsorbed. The *Mariner* close-ups have revealed nothing like our river valleys, although a possible watercourse has been noted in frame 11 (Plate XV). Streams of seasonal character associated with the retreat of the caps in polar regions or with post-volcanic activity may exist, but if so do not seem to have been geologically important for moulding the surface.

Yet, while surface erosion by ice and water will be negligible, underground deposits of ice and possibly phreatic erosion by subsurface water basins may have left a strong mark on Martian landscapes.

The scarcity of surface water is well established by telescopic observation, spectroscopy and the *Mariner* photographs, so that many light soluble compounds which are leached out of our soils and rocks and concentrated in the seas ought to be comparatively abundant on the surface of Mars, all the more so as the planet would have received a higher proportion of the lighter chemical elements at birth (p. 14). Rock salt and other solubles may therefore be important mineralogical constituents of the surface formations, making them highly vulnerable to erosion by water, whether liquid or vapour, wherever it appeared in quantity, as might happen in volcanic eruptions.

The presence of an atmosphere guarantees some subaerial

denudation, and the worn appearance of many Martian rings commented upon in the preceding chapter bears clear witness to it. Frost shattering and exfoliation (heat erosion) may also be expected. Nevertheless, these processes will all be very inefficient as compared with the rapid denudation familar to us. The thinness of the atmosphere and the low gravity will also temper the erosive and abrasive action of the winds, which, moreover, appear to be comparatively slow, not exceeding 30 miles per hour, to judge by the observations of cloud movements. Thus there is no efficient agency to remove the products of areological decay, which will tend to remain close to the place of origin, only the finest particles being entrained by the wind.

Most of our sedimentary rocks, such as shales, limestones, sandstones and conglomerates, have been formed in water, and must perforce be absent from or very rare on Mars. On the other hand, wind-borne or *aeolian* sediments, resembling our desert sandstones and loess, are possible and even probable. Although the rate of production of such sediments will be very slow, they could yet build up to great thicknesses in the course of hundreds of millions of years. Mountain screes could become cemented into breccias. Reckoned in terrestrial terms, such *secondary* formations will be unimportant, the pride of place being occupied by *primary* igneous rocks, not only as basements but on the surface as well. Some metamorphism by heat and pressure cannot be excluded, so that rocks resembling our schists need not be unknown on Mars. They could become exposed in high-fault scarps. Yet thousands of feet of strata have to be removed by erosion before abyssal rocks can outcrop on the surface. The mountains of Mars will not be of granite, and its igneous formations will take chiefly the form of lavas and pyroclastics (ash, lapilli, volcanic bombs), loose or cemented into tuffs, agglomerates and breccias. Under a low barometric pressure lava will tend to bubble up and assume pumiceous texture. Moreover, sialic (acid) lavas are hot and sluggish, tend to cake over rapidly and form bolster-like banks, as in frame 11 of the *Mariner 4* pictures. This will be all the more so on

Mars, as the rate of flow depends on gravity, and the caking-over will be speeded up by the more vigorous expansion of the occluded gases.

The composition of the solid body of Mars in bulk may not differ much from that of the Earth, but the quantitative disparities may become considerable at the two ends of the density scale: in the core and in the upper crust. It is, therefore, quite on the cards that some materials unusual by our standards may appear among the Martian volcanics. Thus, for instance, a volcano on Whakari, an island off New Zealand, erupts molten sulphur and gypsum, instead of conventional magma. A safer guess is that Martian vulcanism will assume extreme acidic forms, including the Peleean type of activity in which huge clouds of hot fluidised ash, known under the French name of *nuées ardantes*, are emitted in violent spasms. Such fluidised ash flows much faster than water, and its flow depends on the agitation of hot gases rather than on gravity, so that it can be equally fast and even faster on Mars. On cooling a *nuée ardante* sets into a kind of rock, often mistaken for lava and called *ignimbrite*, or *Aso lava* in Japan. Ignimbrites probably account for most of the lunar and Martian surface alike.

In these circumstances it seems extremely unlikely that the reddish colorations encountered on Mars are due to limonite or to a coating of limonite on sand grains, which would require water action. For one thing the low degree of oxidation of iron to be expected in Martian conditions would result in blue, not in red, tints; for another, this amount of iron does not appear probable in the topmost layers of the Martian globe. Bauxite, a hydroxide of aluminium, as a source of red colour is more consistent with the segregation of elements according to atomic weight in the primordial nebula (p. 14). It is true that on the Earth it is a product of the weathering of basic rocks in tropical climates, but it could arise in other ways on Mars.

The petrographic picture that emerges from this discussion falls somewhere between the Earth and the Moon, but nearer the latter, with some specifically Martian features present on neither. The greater resemblance to the Moon stems from the scarcity of surface water, which is one of the determining factors of geology. On Earth the friable volcanics erode very rapidly, and, although enormous quantities of them have been formed right on the surface in the geological past, they have been washed away and embedded in sedimentary formations. Only the stumps of the old volcanoes remain, say, in the mountains of Wales. We have no right to say that it never rains on Mars; it may do so sometimes, and a violent volcanic episode, for which there is good observational evidence, as we shall see later, may well result in a real cloudburst. Such developments, however, will be atypical and not a major areological factor. The volcanics and the old volcanic structures will endure long after the cessation of the forces that gave rise to them, and cover up most of the surface.

In combination with a continental-shield type of structure this leads to substantially lunar tectonics, bearing in mind that Martian surface gravity of 2/5 g is much above the lunar 1/6 g. Thus, for instance, while folding, which depends on gravity, is virtually unknown on the Moon, it cannot be ruled out on Mars. All we are entitled to assume is that it will be relatively less effective and will result in fracturing and faulting long before the intricate plications of our strata are reached. By itself, however, without sculpturing by streams and glaciers, surface folding could yield only undulating heights, such as appear in frame 12 in the *Mariner* pictures. Should faulting intervene, abrupt escarpments may arise from this mechanism of relieving contractional stresses of the crust, resulting in landscape forms similar to the Blue Mountains of Australia.

Another mountain form to be expected on Mars is fault-bounded ridges, or *horsts*, and elevated tilted blocks, exemplified on Earth by the Great Basin Ranges of the western U.S.A. or by the Lunar Apennines, although the existence of the latter is closely related to that of Mare Imbrium, which has no Martian counterpart. Indeed, the mountain system shown in frame 13 is an almost straight fault-scarp, modified by cauldron subsidences, and so invites comparison to the Great African Rift, or perhaps the Lunar Altai.

Our alpine ranges, as briefly mentioned on p. 16, owe their origin to the accumulation of water-borne sediments in sea troughs, known as *geosynclines*, their subsequent isostatic sinking, induration by heat and pressure, and eventual extrusion in strong folds by lateral compression arising from crustal shortening. Intrusion of magma at the root of the folds, with or without vulcanism, is a secondary development that appears to have been stronger in earlier orogenies (de Sitter), which seems to indicate a gradual decrease in internal heat owing to the exhaustion of radioactive elements.

It is just possible that wind-borne sediments become packed in an areosyncline without the action of water, eventually to give rise to an alpine orogeny. Such a process, however, would be extremely slow and ineffective, so that this could not happen very often even within the 4½ aeons that have elapsed since the formation of Mars. The developing of a sufficient down-warp in the thick sialic crust of Mars to yield an areosyncline also seems doubtful. To sum up, alpine ranges are very unlikely, if not downright impossible, on Mars.

Such tectonic processes are, of course, not the only way in which mountains are formed. We must not forget igneous activity.

Martian gravity being two-fifths of ours, everything will be so much lighter and easier to lift. We have also gone over the grounds for believing that the topmost formations of Mars are specifically lighter than those of the Earth. In combination this should greatly facilitate intrusion of magma under or between the superjacent rocks in the form of sills, laccoliths, lopoliths and stocks, and, since these rocks are also brittle, the ascent of magma in dykes, sheets or plugs to the surface will occur much more readily than on the Earth. Thus, assuming the same rate of liberation of radioactive heat in the rocks of Mars, intense volcanic activity is to be expected.

Such radioactive elements as thorium, uranium and actinium are very heavy, and so may be thought to have been largely concentrated towards the centre of the solar nebula.

Yet they occur in relatively high amounts in meteorites, assumed to originate beyond the orbit of Mars. Furthermore, these large atoms fit preferentially into the crystal lattices of the acid sialic rocks, so that the radioactive heats liberated in our granite, basalt and ultrabasic dunite respectively are in the ratios of 6·6 : 1·6 : 0·03. Now, since the proportion of sial in Mars is higher than in the Earth, if similar ratios apply, the underground heat of Mars should be likewise higher. Historically the issue is largely hinged on the point in the formation of the planets at which the present association between acid rocks and heavy radioactive elements occurred. If it pre-dates the emergence of the Sun as a star and the subsequent fractionation of materials by volatility within the nebula, the reasoning holds, and the radioactive heat of Mars will exceed that of the Earth.

Yet, speculation apart, we have substantial observational evidence pointing in the same direction.

E. M. Antoniadi was the first to observe in 1910 and 1911 grey circular clouds, which seem to have a point origin, rise rapidly to great heights, estimated at 60–120 miles (Saheki, Murayama), and then spread out mushroomwise in all directions. Four further clouds of this type were reported by Japanese observers in 1950 and 1952. Deucaleonis Regio, a strip of bright ground, somewhat similar to Atlantis, bordering on the very dark elongated area of Sinus Sabaeus, is particularly favoured by these appearances, but on 15 January 1950 Tsuneo Saheki and two other Japanese astronomers observed a similar grey cloud, about 460 miles in diameter, over Erydania and Electris, which are adjacent to Atlantis.

It will be recalled from p. 47 that frame 11 appears to show active volcanoes in Atlantis, and there can be little reasonable doubt that these observations relate to volcanic eruptions of impressive power, although in a thin atmosphere and under low gravity the upsurge of a volcanic cloud will be much faster and its spread wider than on the Earth. These clouds have also been seen to fluctuate in colour between grey, bluish white and dull yellow within the space of a few days, as would be consistent with the volcanic interpretation.

Acidic vulcanism is violent, which fits in well with the views advanced here, but even it falls rather short of accounting for the surprising flares observed on Mars in 1937, 1951, 1954 and 1958, also from Japan, and in 1954 from the U.S.A. These flares appear to be associated with certain localities. Thus three of them were seen in or near Tithonius Lacus (1951 and 1958 observations), and two others on Edom Promontorium.

The flare observed on 8 December 1951 is estimated by Saheki to have endured for five minutes and attained the brilliance of a sixth-magnitude star, which means that, detached from Mars, it could have been seen with the naked eye. The scintillation of the flare may have originated in our atmosphere.

It has been calculated by D. B. MacLaughlin that the famous 'fire fountain' of Vesuvius on 8 August 1779 would have had at the distance of Mars a magnitude of between 16 and 17, and so have been at least 10 000 times fainter than Saheki's flare in Tithonius Lacus. (He saw an even brighter one on Edom Promontorium on 1 July 1954, but it lasted only about 5 seconds.) MacLaughlin goes on to comment that 'if these Martian flares were volcanic, they would indicate that Martian vulcanism is characterised by occasional great outbreaks of incandescent gas a few kilometres in diameter and with temperatures very far above those known in terrestrial vulcanism' (*Sky and Telescope*, February 1955).

This sounds suspiciously like a description of a nuclear explosion. The idea that there may be Martians who were responsible for this, for peaceful or war-like purposes, is not necessarily ridiculous, but it is unwarranted in the present state of our knowledge, and 'natural' explanations are to be preferred.

The enthusiasts of meteoritic impact will, no doubt, contend that bubbles of incandescent gas were produced by large meteorites hitting the surface, but their localisation on Mars militates against this hypothesis. On the other hand, the volcanic interpretation would require a very high subcrustal temperatures and so a concentration of radioactive substances much above the terrestrial. I have, in fact, suggested (1964), somewhat half-heartedly, that, if sufficiently abundant, these substances could become concentrated in veins by the process of mineralisation, thus forming a kind of natural nuclear reactors that occasionally got out of 'control' and blew up.

Without attaching too much importance to this, it may be fairly stated that volcanic activity, seemingly of extremely violent acidic type, is in progress on Mars at the present time, which is in truth hardly surprising and would have to be expected even if no circular clouds or flares had been observed. In these circumstances it is wholly superfluous to have recourse to meteoritic bombardment or any similar extraneous agency to explain the crateriform features of the Martian surface, even if some of them are produced by meteorites, which in our terrestrial experience is an infrequent occurrence of no great geological moment.

A 'crater' as applied to the lunar and Martian ring structures is a somewhat unfortunate name, as on the Earth it commonly denotes the vent of a volcano through which lava, ash, etc. are erupted. No serious student of the Moon has suggested that the lunar craters up to 200 miles across are in fact volcanic *vents*. They are regarded as *calderas* or *cauldron subsidences*. No volcanic vent or crater is known to exceed a diameter of 2½ miles, but a caldera, such as Ngorongoro, may be 15 miles across, while A. R. Crawford discovered in an ancient volcanic complex of southern Australia ring structures certainly 55 and possibly 100 miles in diameter, and the Mogollon Plateau of New Mexico, studied by W. E. Elston (1965), must be put in the same class; it measures 75 miles in width, just as does the large* Martian 'crater' in frame 11.

A caldera may be produced by an explosion when a volcano blows off its top, as did Krakatoa in 1883 and Katmai in 1912, and, since very violent volcanic events have been observed on Mars, it may be assumed that at least some of its ringed depressions have originated in this way. More commonly, however, a caldera is a collapse feature.

It begins with a *laccolithic*, mushroom-shaped intrusion of

* Under lower gravity it could be larger.

magma, which uplifts the overlying rocks into a dome – and a large dome is clearly visible in frame 11. Upward pressure causes the crust to crack in concentric cones tapering towards the head of the intrusion, and liquid magma enters the fractures as so-called *cone sheets*. This behaviour is not limited to magma. Any *incompetent rock*, subject to solid flow, or *rheid* (A. Holmes), can form a laccolithic intrusion and upraise a dome. There are huge salt domes in Iran, while ice also forms domes, known under the Eskimo name of *pingo*, in the permafrost regions of the Earth. Terrestrial pingos do not exceed a few hundred feet in diameter, but on Mars they could attain great size. If the ice evaporates or melts away a caldera will result.

But, to return to magma, it may, of course, set solid, in which case the dome will 'stay put'. Something of this kind has happened in the ancient orogenies within the unfolded continental shields, as exemplified by the Bulawayo system in Rhodesia, which consists of a large number of round granite bosses (*batholiths*, which may be described as 'mountains of magma', not laccoliths in this case). More often than not, however, there is further action. The magma may withdraw from the underground chamber supporting the dome, which then caves in, forming a *cryptovolcanic* structure.

The withdrawal of magma or the emptying of the chamber leads to subsidence and further fracturing of the crust in reverse order, the cones tapering upwards. Once more, magma may become injected into the fractures along which subsidence occurs, and the resulting intrusions are known as *ring dykes*. If the development is sufficiently vigorous and near the surface, the magma in the dykes may reach the surface, leading to annular volcanic activity along the edge of the subsidence. This is why secondary craters appear along the bounding ring walls, and the presence of such craters is, in fact, a sure sign that the caldera is a magmatic cauldron subsidence. This applies to the 75-mile Martian ring in frame 11 and to its 'scalloped' neighbour (Plate XV and p. 47). In this way the bounding walls of the subsidence may be built up gradually by marginal vulcanism. Such a mechanism was proposed for the lunar ring mountains by Loewy and Puiseux towards the end of the 19th century, and Gilbert Fielder has shown (1967) quite conclusively from the *Orbiter* photographs that the walls of the lunar ring Flamsteed P have been formed by the marginal effusion of lava, while my own analysis of the *Ranger 9* close-ups of Alphonsus (1966) similarly indicates marginal emission of volcanic ash, which Elston has also found along the ring fractures of the Mogollon Plateau.

This genesis is peculiar to the largest ring forms, but a collapse of a volcanic cone may also be due to the lowering of the level of magma in the chamber owing to the escape of the occluded volatiles (gases and water) or to eruptive evisceration, and this is a likely mechanism in the case of minor calderas. Finally, as noted on p. 63, pingos may produce calderas as well.

We thus seem to have here a complete answer to the nature and origin of the Martian craters, but the streakiness and the tartan structure of the ground, too, has its counterpart on the Moon. The 'tartan effect' is strongly marked in the *Orbiter* photographs of Copernicus, and the floors of Ptolemaeus and Alphonsus in the *Ranger 9* views are criss-crossed by rows of contiguous small hollows. Fielder has made some detailed studies of the 'grid lattices' in the 'central highlands' of the Moon.

Whether these results are applicable to Mars without modification, I am in some doubt, but at least a proportion of the Martian 'streaks' are undoubtedly crater chains. Others may be tectonic fractures partly concealed by erosion, filled with lava, or packed with debris. The 'canals' would be larger fractures of the same kind, widened by erosion.

7. How Much Air?

That Mars has an atmosphere is immediately clear from the presence of variable polar caps and is further confirmed by the occurrence of mists and clouds. The brilliance of the limb, where surface detail is usually washed out in visual wavelengths (Plate II), is likewise an atmospheric effect. Most ground materials are dark and on the Earth reflect on the average about 10% of the light incident upon them; in other words, their *albedo*, which is the reflected fraction of the incident light taken as unity, is 0·10. Mercury, whose surface is very dark and rough and atmosphere thinner than that of Mars, has an albedo of 0·06, and the Moon one of 0·07. The visual albedo of the Earth is put at 0·36, but most of the reflected light comes from the clouds, which account for 54% of the surface. Without them the Earth would be no brighter than Mars, whose average albedo is 0·16, and without the air not much brighter than the Moon per unit area under equal illumination. Indeed, the air is brighter than the ground, as a single glance towards the horizon will suffice to show, and distant views are usually pale and devoid of colour. This is the same as saying that if we viewed the Earth through a telescope from another planet, the limb would look bright, as the thickness of the atmosphere in the line of sight increases towards it (see also p. 7), and surface detail would be washed out near it.

When the dark hemisphere of Mars shows at the edge, the terminator does not follow the geometrical boundary between light and shade and looks 'fuzzy', the day grading down into night in a *twilight arc* 8° wide, which is estimated to correspond to a scattering atmosphere 20 miles high. A still higher atmosphere has been deduced from the Wright effect, whereby the ultra-violet diameter of the planet is larger than the infra-red (p. 26). In view of the various doubts expressed about the reality of this effect, notably by Kuiper, Wright's original negatives were re-examined in 1965 by R. A. Wells, who was able to confirm an average diameter difference of 3%, although it was occasionally greater. Putting the diameter of Mars at 4200 miles in round figures, this yields a scattering atmosphere 60 miles thick.

Yet even if the Wright effect stems to any extent from the behaviour of the photographic emulsion rather than the 'air' of Mars, it remains unquestionably true that, with the rare exception of the so-called 'blue clearings' – of which more later – the surface of the planet is more or less thoroughly invisible at the blue end of the spectrum, although it is simultaneously clear and distinct in the infra-red, which penetrates atmospheric haze. The variations in the visibility of the surface features are in themselves proof of an appreciable atmosphere.

Indeed, repeated mention has been made of hoar-frost, mists and clouds. All of them, the advance and retreat of the polar caps and the seasonal darkenings and fadings of the maria cause fluctuations in the magnitude and albedo of Mars, so that the quoted figure of 0·16 represents only the mean value of the latter, and when this was measured by Dollfus in 1956, during a widespread veiling of the surface by 'yellow clouds', it went up to 0·235. Nevertheless, the comparatively low albedo indicates a generally cloudless, transparent atmosphere, which is borne out by observation.

On certain assumptions I calculated in 1959 that the average Martian overcast should be about 20%. N. A. Weil notes that on the morning side of the terminator clouds may extend up to the noon meridian, but in the afternoon half of the disc they are seldom seen at more than 45° from the evening terminator. They are also much more frequent and extensive at aphelic than at perihelic apparitions, which all goes to show that they are associated with low atmospheric temperatures (see also p. 28).

The typical white clouds of Mars are thin, with albedoes

between 0·20 and 0·40 (Antoniadi) as compared with 0·78 for our cumuli. Polarimetric analyses by Dollfus show that they reflect light like ice crystals and so must be classed as light cirrus. Antoniadi has found that they occur chiefly in the 'desert' regions, often in flocks of roundish objects, 10–20 miles in diameter. Yet 'brilliantly white sheets' of cloud were seen in 1920, and on 16 November 1958 I observed a white cloud veil spread rapidly over the south polar lands, increasing from a diameter of about 1000 to 2000 miles in 2½ hours. When first seen, it appeared to bulge out above the limb to the west of the pole (Fig. 9). In fact, veils of this kind are not infrequent above the poles, and, unlike the spring and summer appearance of the caps, when they do not project above the limb and are clearly surface deposits, the caps in late autumn and early spring are usually atmospheric and stand up at least 20 miles above the ground.

As a cloud is carried over the limb by rotation it may appear detached from the surface, like the 'defect' in frame 1 (Plate XIII) in the *Mariner* set of photographs (p. 45). Unless the cloud is very large, its height above the ground can then be measured. Altitudes of between 5 and 20 miles have thus been obtained for white clouds.

An important Martian peculiarity is the so-called *violet layer*, which powerfully reflects radiations below 4420 Å and is believed to consist of very fine crystals of water ice and possibly dry ice (frozen CO_2), suspended at a height estimated variously at between 20 and 100 miles (Kuiper, de Vaucouleurs). The violet layer may indeed vary in altitude, as it does in thickness and distribution. When viewed or photographed through a blue or violet filter, Mars often shows a belted appearance, which was first noticed in the photographs taken by E. C. Slipher in 1954. The number of belts is not fixed, but typically there is a bright zone around the equator, giving way to darker belts over the tropical and moderate regions, followed by extensive circumpolar caps.

This structure of the violet layer is the result of the air circulation between the equator and the poles, modified by the Coriolis forces arising from rotation.

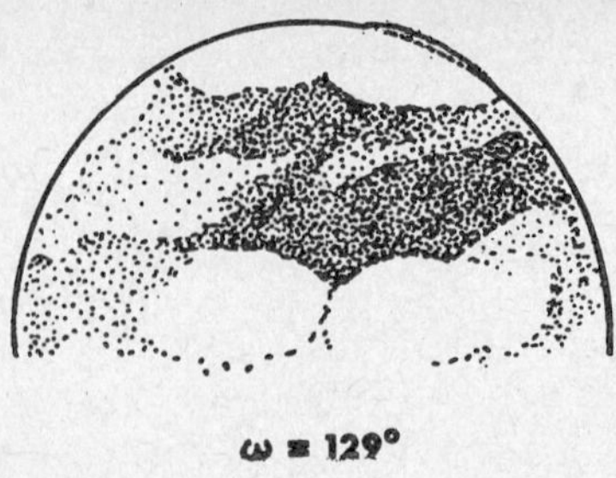

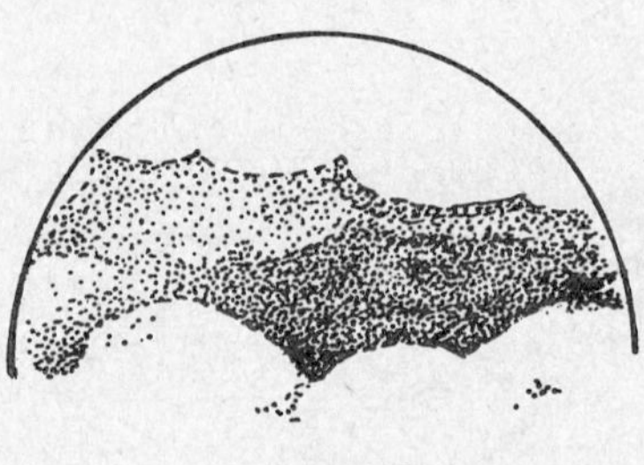

Fig. 9. *Vast cloud, engulfing the south polar lands on 16 Nov. 1958. Observational drawings by the author: the first at 21 h 15 min, U.T., the second at 23 h 30 min U.T. The dark feature is Mare Sirenum, with Mare Cimmerium edging in in the second drawing, owing to rotation. Some of the change in the appearance of the cloud is also due to this, but it can be seen covering areas that were previously clear. This must be ascribed to a general drop in temperature. Note the bulge in the first drawing.* ω = *longitude of the central meridian.*

The relatively warm and moist equatorial air rises to high altitudes, where its moisture is frozen out, forming the bright zone, and continues to flow polewards. Since, though, the linear speed of rotation is greatest at the equator, the equatorial flow is deflected to the west, giving rise to prevailing westerlies and cyclonic weather, as on the Earth. At the same time the relatively dry, cold and heavy air is drawn from the poles towards the equator and experiences an opposite, eastward

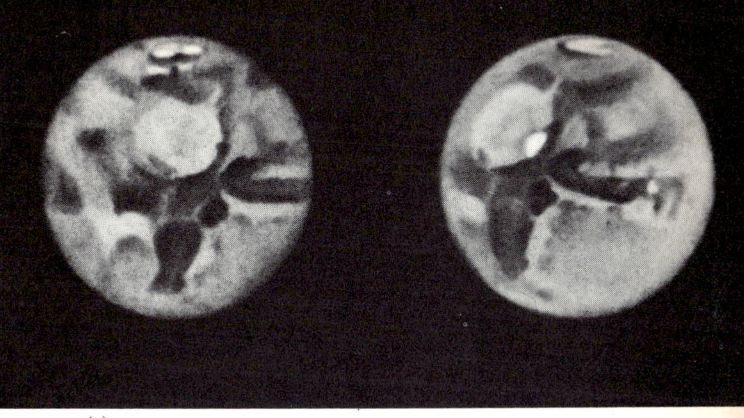

(a)

(b)

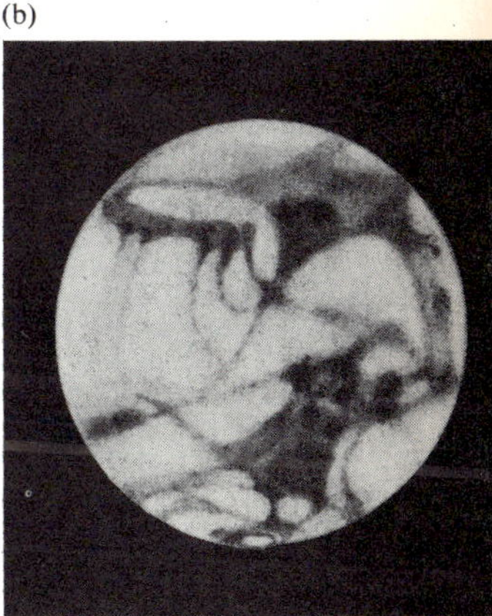

(c)

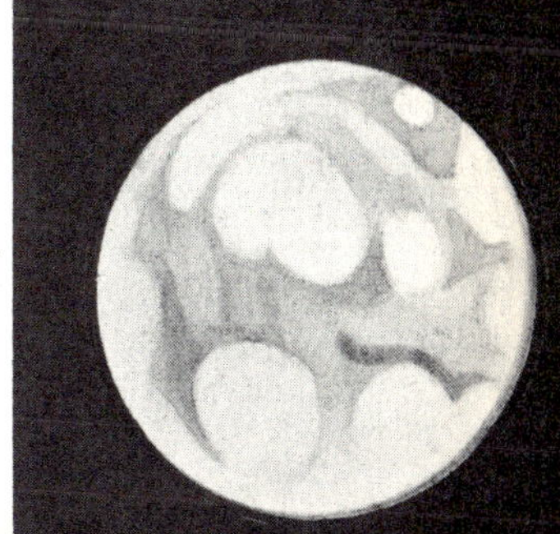

Plate I. *Telescopic appearance of Mars as recorded by different observers.* (a) *The region of Syrtis Major, Hellas and the South Pole by Audoin Dollfus with the 60-cm (24-inch) telescope of the Pic du Midi Observatory on 13 and 14 Sept. 1956, showing the polar cap, white clouds and rotational shift.* (b) *A northern aspect of Mars – Mare Acidalium, Margaritifer Sinus and Sinus Sabaeus – as observed by C. F. Capen on 1 May 1967 with the 82-inch reflector of McDonald Observatory, Texas, and the power of 1000 X. Some 'canals' are shown.* (c) *This observational drawing by the author, made with a 6½-inch reflector on 12 Oct. 1956, red filter and a 'Pola' screen (300 X), comes close to (b) but with a different polar tilt.*

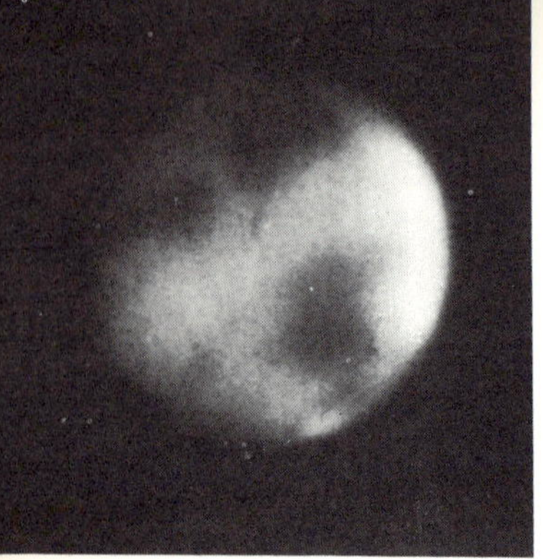

(a)

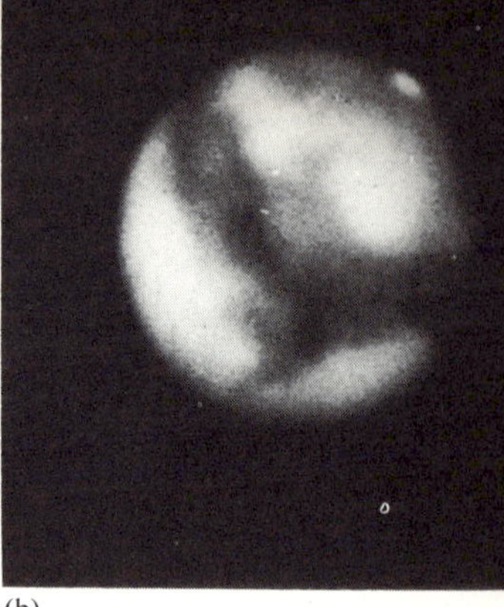

(b)

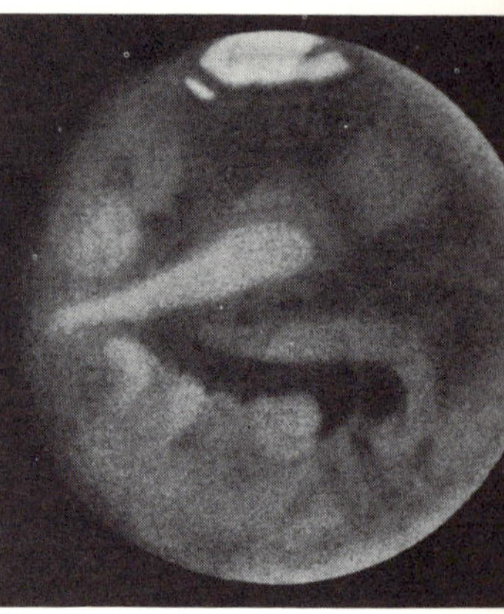

(c)

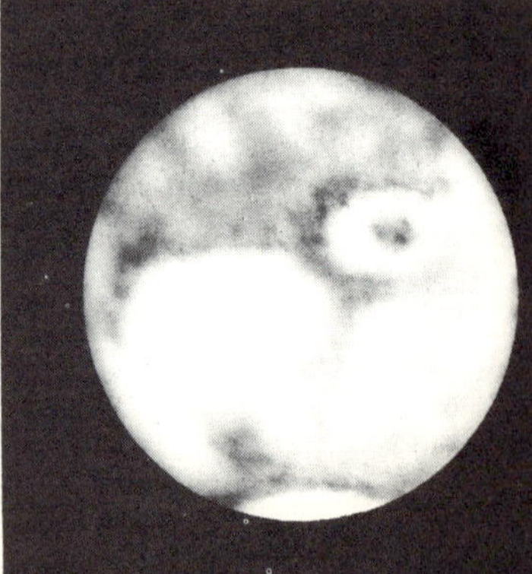

(d)

Plate II. (a) *and* (b) *are telescopic photographs of Mars:* (a) *by C. F. Capen, 29 March 1967, 82-inch McDonald reflector (the original is in colour), shows the north polar cap, Mare Acidalium, Sinus Sabaeus and Mare Erythraeum; an arc of mist is clearly visible along the morning terminator;* (b) *by G. E. Hale, 5 Oct. 1909, 60-inch Lick reflector, shows Mare Sirenum, Syrtis Major and the South Pole.*

(c) *Drawing by Shiro Ebisawa, using the 8-inch refractor of the National Science Museum, Tokyo, marks the discovery of the great yellow cloud over Noachis, above Sinus Sabaeus and to the left of it, 20 August 1956.*

(d) *Mars at 20 hours Universal Time, 27 Dec. 1958, drawn from observation by L. F. Ball, using a 10-inch reflector with a power of 292 X. The main features are the deserts of Xanthe and Amazonis, the north polar cap (bottom), and Mare Erythraeum and Tithonius Lacus above.*

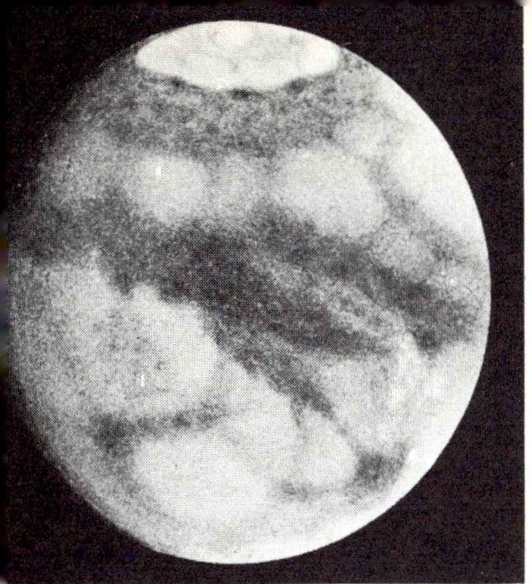

(a)

Plate III. (a) *Mars by Shiro Ebisawa, 28 July 1956, 20-cm (8-inch) refractor, National Science Museum, Tokyo. Note the mottled appearance of the dark areas, also shown in (b). There is a suspicion of crateriform outlines, such as were later discovered by* Mariner 4.

(b)

(b) *By Audoin Dollfus, 18 Sept. 1956, 60-cm refractor, Pic du Midi. Maria Sirenum and Cimmerium are in the central portion.*

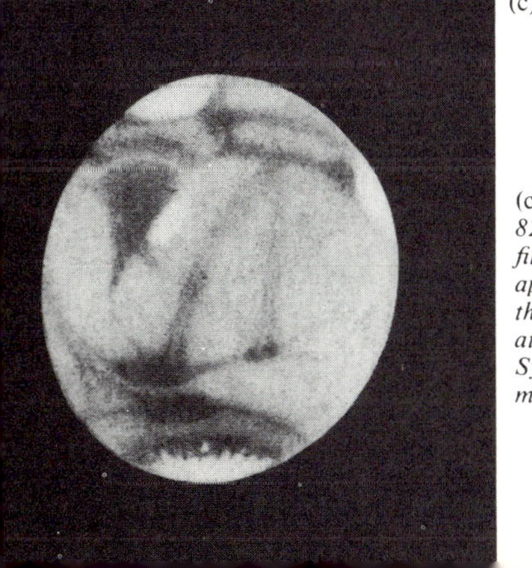

(c)

(c) *By C. F. Capen, 25 Jan. 1967, 82-inch McDonald reflector, orange filter, 1000 X. Note the ragged appearance of the north polar cap, the dark 'collar', 'canals' and clouds at the limb. The dark triangle is Syrtis Major. Mars near its maximum phase effect.*

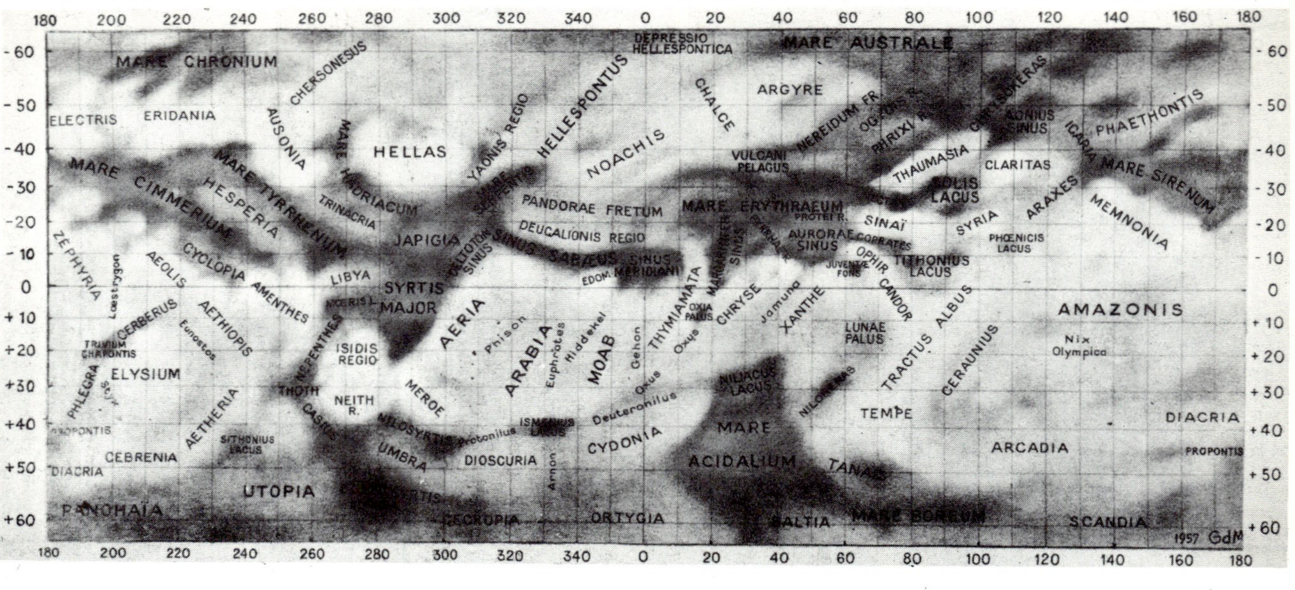

Plate IV. *Mars in Mercator projection – the official map issued by the International Astronomical Union. In accordance with astronomical convention, south is at the top. The map shows no 'canals'.*

VIOLET

INFRA-RED

(c)

(d)

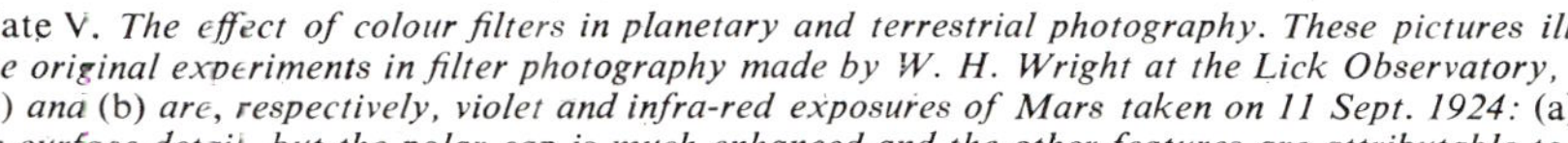

Plate V. *The effect of colour filters in planetary and terrestrial photography. These pictures illustrate the original experiments in filter photography made by W. H. Wright at the Lick Observatory, U.S.A.* (a) *and* (b) *are, respectively, violet and infra-red exposures of Mars taken on 11 Sept. 1924:* (a) *shows no surface detail, but the polar cap is much enhanced and the other features are attributable to atmospheric haze. Syrtis Major and Mare Cimmerium stand out dark and clear in* (b). *This effect is paralleled in* (c) *and* (d), *where the town of San José, 13½ miles from the Observatory, is invisible in violet but is shown clearly, together with the background mountains, in infra-red light.*

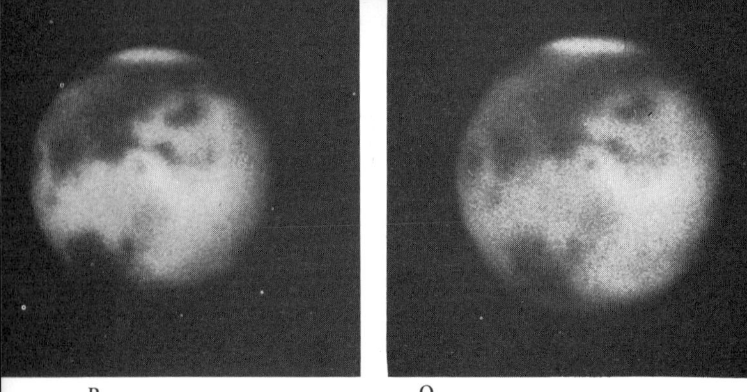

R O

Plate VI. *Progressive filter photographs taken at Lowell Observatory, U.S.A., in July 1954. Red* (R), *orange* (O), *yellow* (Y) *and blue* (B) *filters were used. The maria are clearest in the red, somewhat paler in the yellow (where, though, there is more detail in the deserts) and disappear altogether in the blue exposure. The polar cap appears larger in the opposite order, and the blue photograph shows traces of a belted arrangement in the violet layer, which it mainly portrays. (Lowell Observatory photograph.)*

Y B

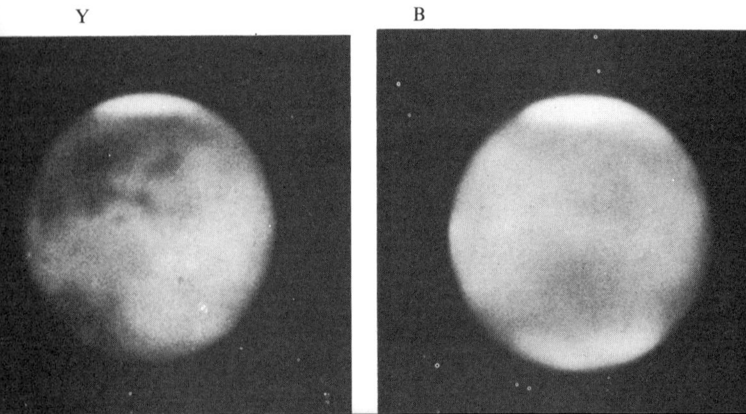

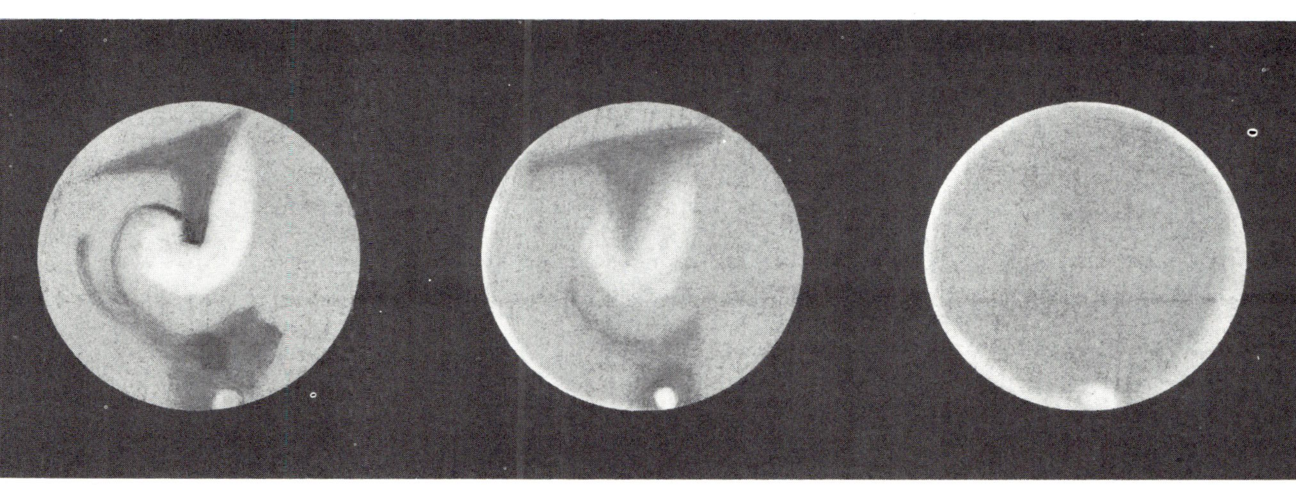

Dufay RED — Dufay GREEN — Dufay BLUE

Plate VII. *This is a visual sequence analogous to the photographic one in Plate VI. The drawings were executed by Alan W. Heath on 9 March 1965, using a 12-inch reflector with a power of 318 X in good seeing and a tricolour separation set of filters: the Dufay red, green and blue, from* left to right. *Note the brightening-up of the east limb, due to morning haze, in green light, and the general brightness of the limb, which is an atmospheric effect, in blue.*

Plate VIII. *A map of Mars in Mercator projection, compiled at Lowell Observatory in Percival Lowell's lifetime, showing* **numerous** *'canals' and 'oases' as thin spidery lines and regular round blobs respectively. (Lowell Observatory photograph.)*

(a)

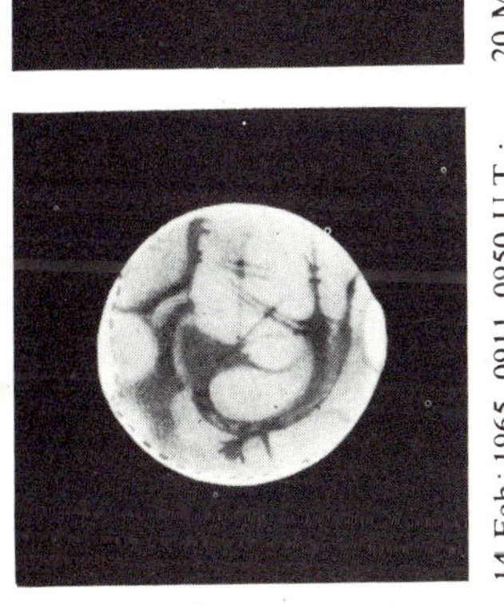

14 Feb. 1965, 0911–0950 U.T.; C.M. 294°; M.D. 31 May; 84-inch reflector; 500X.

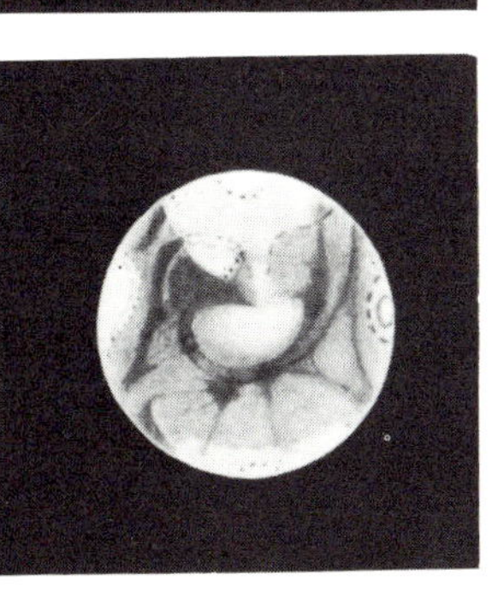

20 Mar. 1965, 0340–0525 U.T.; C.M. 273°; M.D. 16 June; 30-inch refl.; 450X and 800X.

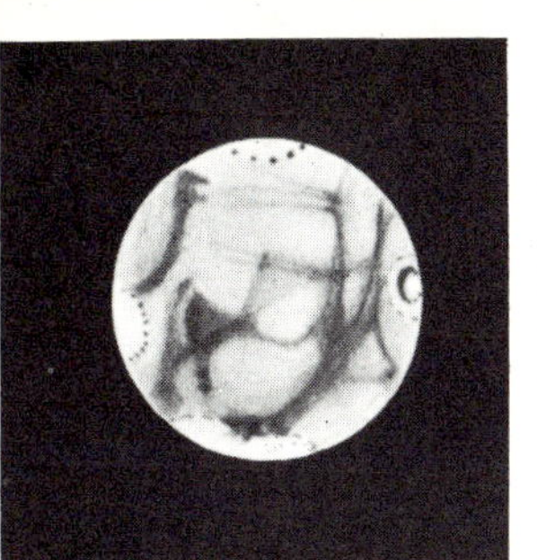

21 Mar. 1965, 0645–0740 U.T.; C.M. 303°; M.D. 16 June; 30-inch refl.; 400X and 500X.

(b)

3 May 1967, 1500 U.T.; C.M. 197°.

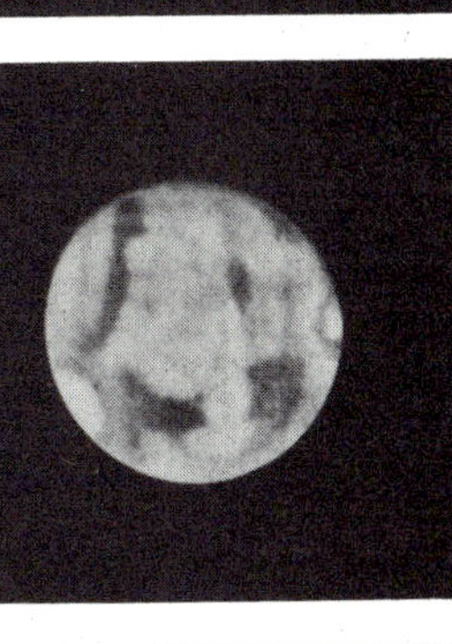

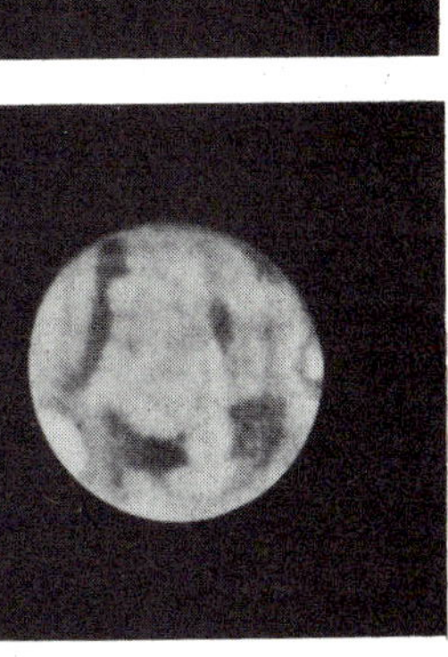

18 May 1967, 1130 U.T.; C.M. 4°.

23 May 1967, 1145 U.T.; C.M. 322°.

Plate IX. *Canal-like markings as seen by modern observers:* (a) *by C. F. Capen in the U.S.A. during the 1965 opposition, and* (b) *by Shiro Ebisawa in Japan during the 1967 opposition period. (Ebisawa used powers of 400 X, 450 X and 600 X on the 12-inch refractor of Kwasan Observatory and the 12-inch reflector of Murayama Observatory.)* U.T. = *Universal Time,* C.M. = *Central Meridian,* M.D. = *Martian Date.*

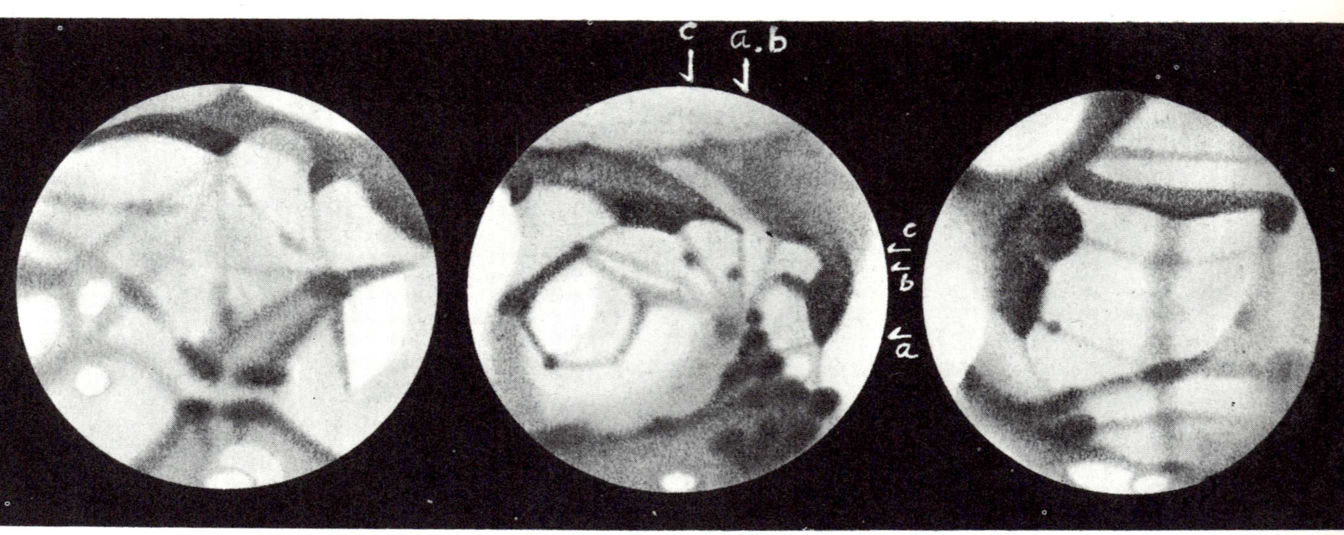

Plate X. *Mars during the latest apparition, April–May 1967; drawings by Tsuneo Saheki, using a 20-cm (8-inch) reflector and powers of 330 X and 400 X. From* left *to* right: *2 May, 13 h U.T.; 25 April, 14 h 10 min U.T.; 22 May, 11 h U.T. Some canal-like markings, round white clouds (extreme left) and rounded dark spots are shown.*

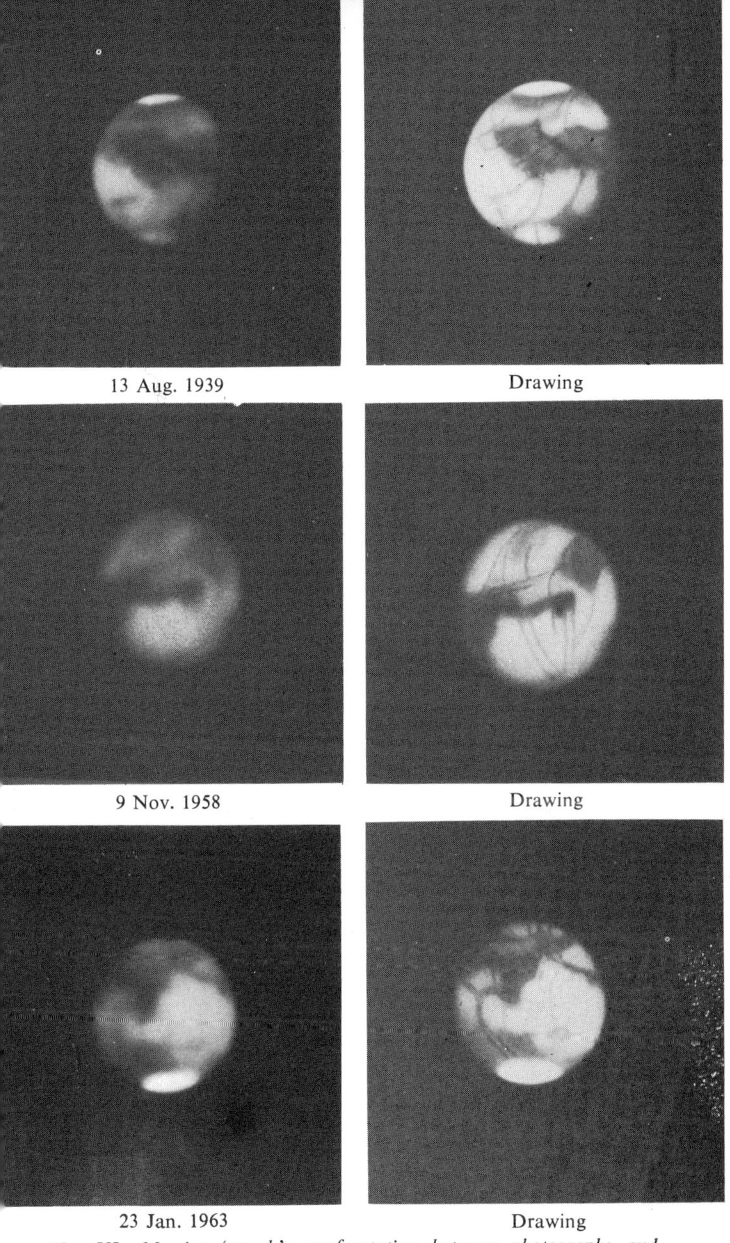

Plate XI. *Martian 'canals': confrontation between photographs and drawings, 1939–63. See text for particulars. (Lowell Observatory photograph.)*

(a)

Plate XII. (a) *The 'eruptive cloud' of 15 Jan. 1950, drawn by its discoverer, Tsuneo Saheki, at 19 h 30 min U.T. (20-cm reflector, 400 X).*
(b) *Two drawings of Mars by Audoin Dollfus, 29 Aug. and 30 Sept. 1956, showing rapid changes in the region of Solis Lacus and the South Pole. Such changes are difficult to contemplate without an adequate atmosphere. (60-cm refractor, Pic du Midi.)*

(b)

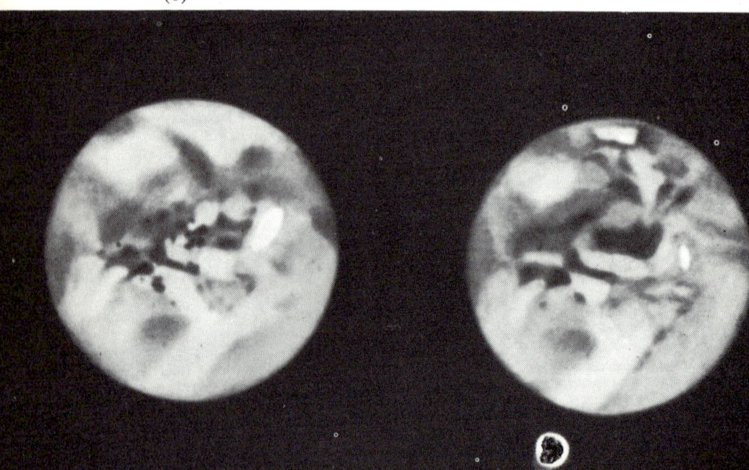

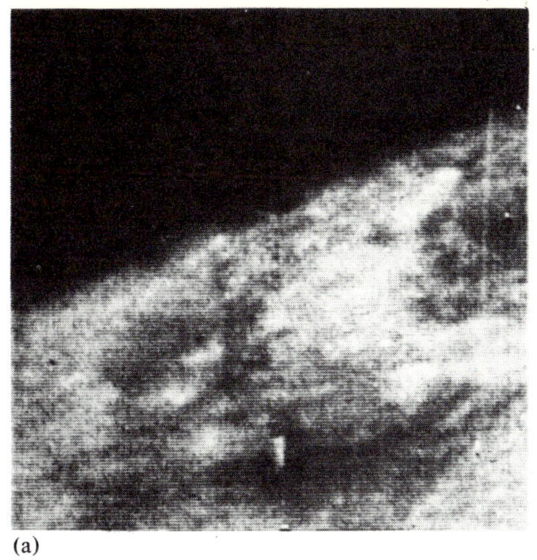

(a)

Plate XIII. Mariner 4, *frame 1:* (a) *has been specially processed to intensify contrast, which is largely lost in the bright lower view* (b). *On the other hand, the latter appears to show light cloud above the horizon. The bright detail is most probably all cloud, but there is a suspicion of rectilinear canal-like features. Refer to Plate IV (map). (Courtesy: U.S. Information Service.)*

(b)

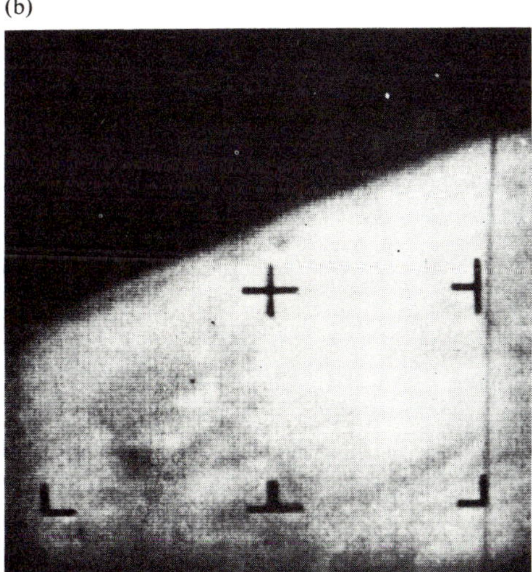

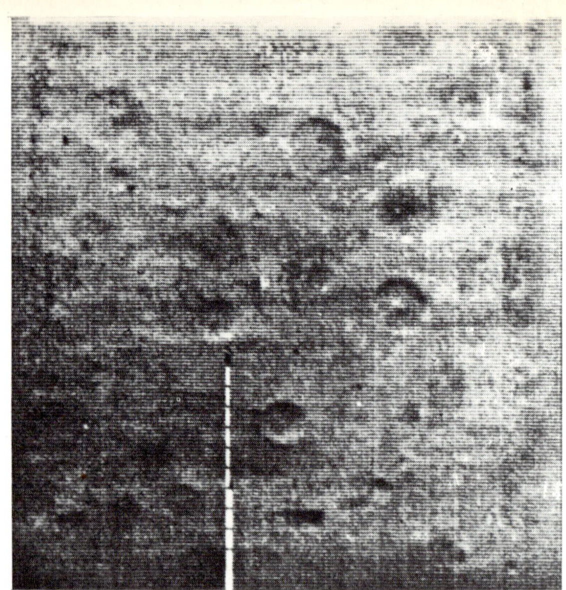

Plate XIV.

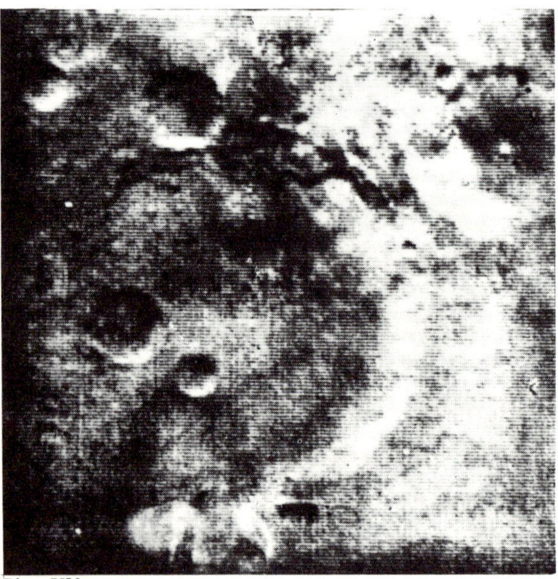

Plate XV.

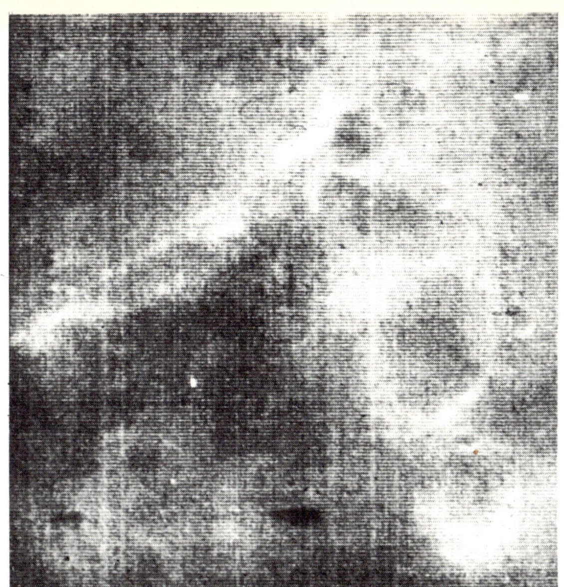

Plate XVI.

Plate XIV. Mariner 4, *frame 9, taken through an orange filter covers about 170 miles in the east–west direction and 160 miles in the north–south direction, and shows numerous shallow crateriform structures, as well as faint irregular features which could be vegetation, in the region of Mare Sirenum bordering on Atlantis. (Courtesy: U.S. Information Service.)*

Plate XV. Mariner 4, *frame 11, reveals a pseudo-lunar landscape, with a 75-mile walled plain and several smaller craters, some deeply trenched and scalloped. (Courtesy: U.S. Information Service.)*

Plate XVI. Mariner 4, *frame 13, shows an area of 170 × 180 miles on the border of Mare Cimmerium and Phaethontis, including a mountain escarpment attaining 13 000 ft above the ground at its foot, and more craters outlined dimly by hoar-frost in increasing atmospheric haze. (Courtesy: U.S. Information Service.)*

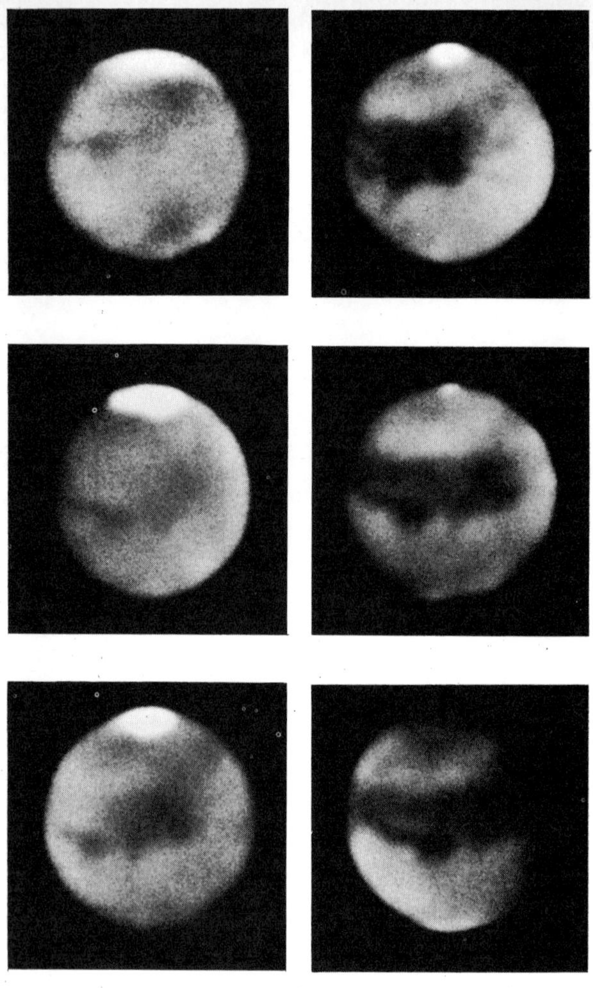

Plate XVII. *The march of seasons. As the south polar cap dwindles, the maria darken and expand. The sequence is from* left down *to* right down. *(Lowell Observatory photograph.)*

thickening of the layer over the poles (see Plate XVII), all clearly show that water ice is the main substance involved. But at the low temperatures of the high atmosphere of Mars some crystallisation of CO_2 is also to be expected, as mentioned on p. 67.

The critical temperature of CO_2 is 31·1°, so that its condition on Mars is that of vapour. It freezes at −65°C, but its boiling point at 760 torr is −78·2°C, going down to −108·6°C at 40 torr and to −134·3°C at 1 torr which shows that its vapour pressures are high. Indeed, the pressure curve only bends down sharply below −150°C. This means that CO_2 cannot make an appreciable contribution to at least the lower part of the violet layer, expected to coincide with the Martian tropopause (marking the upper limit of ascending currents). Yet something more than water ice is needed to account for the total effect, because water ice is a good reflector of ultra-violet rays, while the ultra-violet albedo of Mars is very low, ranging from 0·04 to 0·08.

This led F. Link (1950) to suggest that the layer consisted of or contained meteoric dust, while E. J. Öpik revived an earlier idea that it was a fog of carbon particles – in fact, smoke, no plausible source of which, however, can be found. Volcanic dust has likewise been proposed, which, though, could hardly provide a permanent feature of the Martian atmosphere.

Yet this atmosphere appears to contain a substance that ought to satisfy the requirements to the hilt. This is ozone (see p. 20). It strongly absorbs the ultra-violet and is of a deep blue colour when liquefied. Its critical temperature is −12·1°C, so that it, too, is a vapour at least in the upper atmosphere of Mars. The 760-torr boiling point will be lowered appreciably at the level of the violet layer. Unlike CO_2, however, O_3 has low vapour pressures. It boils at −163·2°C under 10 torr, at −180·4°C under 1 torr, but its vapour pressures drop rapidly beyond this point, being only 0·11 torr at −182·96°C. No experimental data are available for still lower temperatures, but what is relevant for condensation is not just the overall partial pressure of the vapour in the supernatant atmosphere, but its local concentration, and if ozone were formed locally by the action of the ultra-violet radiation of the Sun it could easily exceed, say, within a blue cloud, the critical concentration needed for liquefaction to occur, even if the partial vapour pressure were too low. Moreover, it will be energetically adsorbed by water and CO_2 crystals, which will become coated with ozone molecules. F. S. Johnson (1965) puts the temperature at 100 km (60 miles) above the surface of Mars at 85°K = −188°C, which ought to be wholly adequate for condensation to occur.

If this interpretation is correct the structure of the violet layer may be much more complex than is generally assumed. There would be a lower layer, at the tropopause, consisting chiefly or solely of water ice crystals; this would be overlain by a thin dusting of CO_2 haze; and at 50–60 miles above the surface would stretch the violet layer proper, consisting of a fog of liquid ozone. All three should thicken at nightfall, both owing to the lowering of atmospheric temperatures and of their concommitant altitude, bringing them into a denser part of the atmosphere where more relevant material is available.

At or near the ground level a 'glow of vaporisation' has been observed along the morning terminator, and Camichel's polarimetric investigation shows that it is due to water droplets about 2μ in diameter, so that water fog is possible on Mars. As we have just seen, however, its air is very cold; and although such supercooled droplets occur in the summits of our cumuli at temperatures as low as −39°C, it is doubtful if anything resembling our convective clouds is known on Mars.

On the other hand, one of its peculiarities is 'yellow clouds' of uncertain status. They are much more extensive and enduring than the other types, sometimes obscuring millions of square miles for days and weeks on end, as they did in 1924 and 1956. It is usual to refer to them as 'dust-storms' – on the inspiration, it seems, of the American Dust Bowl. But this interpretation is bound up with some difficulties.

In 1964 J. A. Ryan in the U.S.A. investigated the dynamics

of the situation, with the result that the observed wind velocities would be adequate to raise dust particles into the Martian air if the ground barometric pressure were 80 mb or more, but would fall short of the requirements if it were 25 mb or less, while the still lower pressures deduced from the *Mariner* 'occultation experiment' (p. 50) would completely rule out any dust-storm on Mars. They would seem to rule out most of the other observations as well, but this is another story.

Antoniadi observed a yellow cloud on the limb completely detached from the ground, so that it could not be simply dust whipped up from the surface by the wind. Indeed, the diligent student of Mars E. C. Slipher at Lowell puts the altitude of the yellow clouds at 18–20 miles, which is substantially the same as found for white clouds. Finally, B. Lyot in France has found that the interposition of a yellow cloud cuts down the polarisation by the ground, which would not be the case if the latter were covered with the same dust. Perhaps, though, we need not take the dust hypothesis too seriously in any case.

Alternatively it has been suggested that the yellow clouds may be of volcanic origin. Owing to the lower gravity, the *scale height* of the Martian atmosphere is greater, so that on a rough reckoning, assuming the same composition and temperature (neither of which is true), it is necessary to climb 2·68 times higher on Mars for the starting air density to drop to a half. For this reason, although the barometric pressure at ground level may be a small fraction of one atmosphere, the Martian stratosphere will yet be both higher and denser than the terrestrial. It is also true that fine volcanic dust remained suspended for weeks and months in the upper atmosphere of the Earth after the great blow-ups of Krakatoa in 1883 and of Katmai in 1912, having been conveyed there not by ordinary ascending air currents, but by the force of the eruption. On Mars this mechanism would be even more effective. Violent volcanic activity has been observed (p. 60f.), so that the explanation is feasible. Yet there is no evidence that the yellow clouds were in any way connected with these events, which must be infrequent, whereas the clouds are not.

Be this as it may, dust of any provenance could provide nuclei for the condensation of atmospheric moisture, so that these clouds could be of a mixed nature – partly dust and partly water (ice). This was the conclusion reached by E. C. Slipher from the observations made during the 1956 apparition, when some desert areas were seen to darken temporarily after having been covered by clouds, including yellow clouds.

On the ordinary terrestrial reckoning one would interpret this as a fall of rain. But the air of Mars is too thin and too cold for that in ordinary circumstances. On the other hand, the ground in summer is quite warm (see next chapter), and a light snow shower would soon melt on reaching it and could produce a darkening, with the possible temporary growth of vegetation, as happens in our arid regions. We have the observations, but the rest is theory, based on uncertain assumptions, on which it is unwise to pin too much faith.

Crucial to this and many other meteorological issues, as well as the existence and nature of Martian life, are the composition, total mass, ground density and temperature of the atmosphere. In theory – on the laboratory level – all, or nearly all, of these could be determined by spectral analysis alone. The constituents of a gaseous envelope reveal themselves as absorption lines or bands in the spectrum of the reflected sunlight; the width or intensity of an absorption line indicates the amount of the gas in the line of sight; temperatures and pressures can be inferred from the molecular rotation-vibration absorptions; and the various results can be checked up against one another. The multiplex interferometric spectroscopy, originated in England by Peter Fellget, has been evolved by him, Janine and Pierre Connes in France, and others since 1949 into a very powerful tool for the study of the longer infra-red radiation, whose energy is too low for high dispersion.

The essence of the method consists in isolating a part of the spectrum under study as it were *en bloc* and feeding it as a single collected beam into an interferometer. The latter is a device comprising half-silvered and fully-reflecting mirrors which split the beam into two equal parts, the path of one being fixed, while that of the other can be modified at will by means of a movable mirror. When the two partial beams are brought together again, the radiation of any given wavelength within the two will be in counter-phase if the movable mirror is appropriately set. This radiation will therefore be extinguished by interference, and as the movable mirror is moved along, which can be done with great precision by electromagnetic means, a black line scans the spectrum. Thus what the detector receives and records is not the weak emission at a single wavelength, but the total energy of the selected spectral sector *minus* one wavelength. This description is oversimplified, as in reality the other wavelengths are also affected, but the total pattern is predictable and susceptible of mathematical analysis. The resulting record is a kind of negative, where absorption lines appear as pinnacles and regions of high radiation as valleys, superimposed on a sufficiently high background energy to ensure an adequate response from the detector. Accidental fluctuations can be smoothed out by repeated scans or by lengthening the time of scanning. The record is subsequently unscrambled into an ordinary spectrum graph by a computer.

Extraordinarily low concentrations of scarce gases can be detected in this way, a point to which we shall have occasion to return in Chapter 9.

Yet it is not all plain sailing; for instance, it has been mooted that the very strong infra-red absorptions of carbon dioxide in the spectrum of Venus are due to multiple scattering in the Venusian clouds. Under low pressures and temperatures, absorptions become thin and spidery. The reflection need not necessarily originate at the surface, which leads to errors in the assumed light path, and so on.

The greatest difficulties, however, arise from the intervention of our own air, and its ozone and water vapour in particular. Ozone completely cuts off from the ground the ultra-violet rays beyond 2900 Å; water vapour similarly blankets the infra-red, except for the 'water window' between 8 and 13μ (1μ or micron = 0·001 mm). It so happens that the normal absorptions of such common gases as nitrogen, hydrogen, helium and argon lie in the region of the spectrum blocked out by ozone. Water vapour eliminates most of the 'planetary heat' in the lower infra-red.

In addition to these limitations, our air superimposes its own *telluric* absorptions on those of the planetary spectrum. Such gases as are absent from or form a minor proportion of our air, as is the case of carbon dioxide, will therefore be readily revealed; but the detection of oxygen or water vapour becomes very difficult, as the corresponding planetary and telluric absorptions overlap. Water vapour is concentrated in the lower atmosphere and frozen out at the *cold trap* of the tropopause, which eliminates it almost wholly from the stratosphere. This makes it possible to measure weak water absorptions due to other planets from stratospheric balloons. The case of the other gases is far more difficult, and the only sure way of escape lies in observation from outside the Earth's atmosphere. This is wholly feasible in the present state of space technology, but has not been successfully accomplished to date.

There exist, however, two conventional methods of tackling the problem.

In one, the spectrum of a planet is compared with that of the Moon at the same altitude and in the same atmospheric condition, especially as regards humidity. The Moon serves as an airless standard of reference, and anything that does not appear in the lunar but does in the planetary spectrum will be due to the planet's atmosphere. The method is not very sensitive.

The other method relies on the Doppler shift arising from the relative radial motion of the Earth and the planet. With the planet approaching the Earth, all its spectral features will be shifted towards the violet end of the spectrum, and vice versa in the case of recession. Since, though, planetary motions are substantially parallel, the resulting shifts are insufficient to separate the absorption lines; the most one can hope for is a relative broadening of the lines and variation in the 'wings' of the lines. With high-dispersion spectra and various technical improvements in recent years this approach has not been wholly unfruitful, but the numerical estimates based on the results thus obtained, whether positive or negative, do not deserve much confidence, as subsequent experience has amply demonstrated.

For a long time the atmosphere of Mars yielded no indication of its composition. Faint absorptions by oxygen and water vapour were claimed (Adams and St John) and denied, and various upper limits to the possible amounts of these were set, above which, if present they ought to have been detected. The first breakthrough came in 1947, when G. P. Kuiper identified CO_2 in an amount estimated by him at twice that in our air, using the Moon comparison method. A year after, the same investigator traced absorptions due to water ice at low temperatures in the polar caps of Mars, thus disposing of the earlier hypothesis that the caps were deposits of frozen carbon dioxide.

Further absorptions of both water vapour and carbon dioxide have since been found by various workers and methods. Their estimated amounts differ, but their presence in the atmosphere of Mars may be regarded as beyond dispute. On the other hand, the early reports of a strengthening of the telluric oxygen lines in the Martian spectrum remain unconfirmed. Even so, CO_2 is dissociated by the short ultra-violet below 1692 Å to CO and O, so that there must be at least some oxygen in the upper atmosphere from this source. This process should also lead to the formation of ozone (O_3), which is important inasmuch as it provides a shield for organic molecules of our pattern against the disruptive ultra-violet radiations. Even on the rather unfavourable assumptions made by C. Prabhakara and J. S. Hogan, Jr (1965) the amount of ozone would be about 0·1 of the terrestrial (approximately 0·3 cm-atm). These investigators place its maximum concentration close to the surface, where it would cause some atmospheric heating.

In 1960 C. C. Kiess, S. Karrer and H. K. Kiess reported nitrogen oxides, NO_2 and N_2O_4, with NO and N_2O_3 as transitional forms, and even sought to explain the seasonal colour changes on this basis. The later search for these compounds (Sinton, Owen) has failed, however, so that, if present at all, their concentrations must be very low, because their telluric absorptions are too weak to block off a contribution from Mars.

From the study of 8689 Å and 10 488 Å infra-red bands of CO_2, T. Owen derived in 1966 a Martian abundance of this gas of 65 $\pm$ 20 equivalent metre-atmospheres; while H. Spinrad *et al.* put it even higher, at 90 metres, which would correspond to a partial pressure of 6·6 mb at the surface, from the examination of the high-dispersion infra-red spectra obtained at McDonald and Lick Observatories.

In Russia A. K. Suslov estimated (1966) the mean water-vapour content of the Martian air at 0·02 g/cm^2 (of surface). A. Dollfus's earlier work (1964), involving observations from Jungfrau Pass (3·5 km = 2·2 miles) in the Alps and from a stratospheric balloon at 14 km (7·5 ml), gave a similar, if slightly lower, figure of 0·015 g/cm^2 from the 1·4μ band. The latest published result (1967) comes from R. A. Schorn, H. Spinrad and collaborators in the U.S.A. It is based once more on the high-dispersion spectrograms taken at McDonald and Lick during the 1964–65 apparition and the Doppler-shifted rotational 8200 Å line, which shows considerable variability in the atmospheric content of water vapour with time and place, this being predictably highest near the poles. The average comes out at 10–20μ of precipitable vapour in rough agreement with the measurements made on 2 March 1963 by the unmanned balloon *Stratoscope 2* at 77 000 ft in the Martian infra-red between 1 and 7μ. This means that if all this vapour were condensed it would yield a vanishingly thin film of moisture, 0·01–0·02 mm thick, covering the surface of the planet. However, Dollfus, observing at Jungfraujoch in January 1963 through very dry frosty air of the Alps, found a figure of 0·2 mm, or at least 10 times as much as the Americans.

No great weight need be attached to any of these estimates, but it remains quite clear that the atmosphere of Mars is cold and dry.

Its total mass was put by Dollfus (1952) at about one-fifth of the terrestrial from polarimetric data, and on the assumption that it had the same composition as our air he obtained a ground barometric pressure of 83 mb (1 atmosphere = 1033 mb). Values between 80 and 100 mb were generally accepted until recently, and the 1963 edition of C. W. Allen's *Astrophysical Quantities* proposes for Mars the following atmospheric mixture:

Ar	N_2	O_2	CO_2	H_2O Vapour	N_2O_4	
2000	178 000	<200	420	10	<1	cm-atmospheres

Argon and nitrogen being undetectable, the corresponding figures are pure guesswork, to make up the balance. We have seen on p. 80 that the actual amount of CO_2 may be as high as 9000 cm. On the other hand, an opinion has lately gained ground that Dollfus's figure for ground-level pressure may be too high, owing to his not having taken into account the scattering and polarisation by Martian aerosols.

Indeed, thin as the air of Mars may be, it has high 'optical opacity', and our views of the surface are mellowed by what Kuiper has described as a low-lying 'fair-weather haze' and Kozyrev as a 'yellow-green haze', which according to him cuts off the wavelengths below 5000 Å. The comparative softness of shadows in the *Mariner* photographs has already been remarked upon (p. 48). These show increasing haziness as the morning terminator is approached, and all surface detail is lost in the last six frames.

Yet a recent investigation (1966) by A. Dollfus and J. Focas indicates that the air of Mars scatters light according to Rayleigh's law for pure gases, the scattering coefficient being inversely proportional to the fourth power of the wavelength. This would mean that the air is clear and substantially free from suspensions. A ground-level barometric pressure of 30 mb is inferred by Dollfus.

By a different method, from the pressure broadening of the CO_2 absorption bands, Kaplan and collaborators (1964) deduced an atmospheric pressure of between 10 and 40 mb.

But this is all a matter of theoretical models. In the models corresponding to the lower figure it is assumed that either there is no nitrogen and argon at all, or that their respective partial pressures are 2 mb each, while that of CO_2 is 6 mb. Similar figures have been proposed by some other workers.

J. W. Chamberlain and D. M. Hunten have, however, subjected the various methods used for calculating the pressure and mass of the Martian atmosphere to rigorous analysis (1965). They conclude that the polarimetric and photometric techniques are not nearly so reliable as has been supposed, while the spectroscopic approach, though fully adequate in theory, cannot at present settle the issue of either the total atmospheric pressure or even the partial pressure of CO_2. R. M. Goody's discussion also stresses the difficulty of obtaining reliable numerical results.

This applies even more to the *Mariner* 'occultation experiment', to which some of the other recent estimates have been tailored, because the result involves some unverifiable assumptions regarding the structure of the ionosphere of Mars. The calculated ground-level pressures of 5–10 mb would leave Mars with an atmosphere of substantially pure carbon dioxide, with some dissociation to CO and O at higher levels, and this appears highly improbable if what is known about volcanic exhalation is taken into account. On the Earth, CO_2 forms only 0·03–0·05% of the atmospheric mixture.

At 5 mb and even 10 mb the sky of Mars would look black, except towards the horizon. The density of the shadows depends on the brightness of the sky, so that if the sky of Mars were really black the shadows would be very dense and dark, as on the Moon. Yet the *Mariner* photographs show them pale and diffuse, which necessitates a bright sky. Indeed, the totality of observational data – clouds, mists and hazes, hoar-frost and polar caps – appears incompatible with an atmosphere of such extreme tenuity.

As that pillar of English commonsense, Dr Samuel Johnson, said, 'There can be no security in the conclusions if the premises are not understood.'

8. Climate

Estimating the climate of an inaccessible place, such as Mars, is not an easy task. The soft-landing of an instrumented *Voyager* probe is scheduled by the N.A.S.A. for 1971, and there have been plans to modify the lunar *Apollo* craft for carrying a seven-man expedition to Mars by 1984, in which the powerful *Saturn 5* would be used as launcher. All this is within the 'capability' of the present-day space technology; so far as the manned mission is concerned the Moon has to be reached first and much will depend on the continued success of the *Apollo* venture.

Thus it is still a long way to Mars, and for the time being we have to rely on distant observation.

Since Mars is a terrestrial planet, the Earth forms a natural control, a standard of reference for testing the fitness of hypotheses and correlating the findings about this planet. In most cases, however, proper comparison involves viewing our home planet from space, to see what it would look like to an outside observer using the same type of instruments as we do in studying the other worlds. Indeed, it is a matter for surprise that, while so many probes, orbital satellites and other gadgets are being shot yearly beyond our atmosphere, no attempt has been made to obtain the comparatively easy spectroscopic, radiometric and other relevant space data about the Earth, which could be of great assistance in evaluating the information gained by much painstaking research and cutting out the dead wood of useless controversy. Burns' advice 'To see oorsels as ithers see us!' seems to have fallen on deaf ears.

Be this as it may, we must attempt to strike a balance between the meagre body of incontrovertible fact, surmise and the inherent limitations of our research 'weaponry'.

At the Earth's orbit the energy of sunlight passing through a square centimetre exposed normally to it within one minute (solar flux) suffices on the average to raise the temperature of one gramme of pure water by 2°C. This is, of course, the Solar constant (p. 12). At the orbit of Mars this energy is reduced by a factor of 0·52 at perihelion and 0·36 at aphelion, the mean allowance being 0·43 of the terrestrial, or 0·86 cal.

From this it follows clearly that Mars may be expected to be colder than the Earth. But before we jump to any unwarranted conclusions, let us consider that the energy of sunlight incident upon a unit area of the Earth's surface is, with the Sun at the equator, proportional to the cosine of the geographical latitude. This means that at perihelion the equatorial region of Mars will receive, assuming equal atmospheric obscuration, the same amount of solar heat as a point on the Earth at a latitude of about 59° at equinox. This is the latitude of the Orkneys or southern Scandinavia, which may at this time enjoy quite genial weather. At aphelion the corresponding latitude will shift to 69°, which is, admittedly, beyond the Polar Circle, but still not deadly to life 'as we know it'. In fact, the skies of Mars being usually clear and the atmosphere transparent, it will be rather warmer there in daytime than the terrestrial correspondence might indicate.

On the laboratory level, if Mars were a *perfectly black body* and so behaved like a hole in an airless hollow sphere internally coated with soot, it would absorb all of the solar energy incident upon it in proportion to its cross-sectional area. It would also obey the Stefan-Boltzmann equation $E = \sigma T^4$, where E is the received energy, σ is Stefan's constant $= 5{\cdot}75 \times 10^{-5}$ erg/cm^2/sec/deg^{-4}, and T the absolute temperature, referred to as *black-body temperature*. Since Mars is a sphere in rotation, the absorbed energy would be distributed (unevenly in time and space) over its entire superficies, which entails an average reduction by 4; and, since T is proportional to $\sqrt[4]{E}$, the corresponding reduction in absolute temperature will be by a factor of $\sqrt[4]{4} = \sqrt{2} = 1{\cdot}41\ldots$. The temperature thus computed would be the true mean annual temperature of a perfectly black airless sphere and is known as *blacksphere temperature.*

It is clear, however, that the fraction of the sunlight reflected back by the planet cannot contribute to the heating of its surface. This fraction is its albedo (p. 65), but if we are to be accurate we must take not the visual albedo, but an albedo extended to invisible radiations as measured by a *bolometer* or other energy-sensing device and called *bolo-meteric albedo*, which in the case of a planet is usually somewhat lower than visual.

If we now substitute the absorbed part of the energy left over by the bolometric albedo for the total energy E in the Stefan-Boltzmann law we will obtain the *grey-body* and *grey-sphere* temperatures of the planet, which form a more realistic approximation to the actual conditions. It is still assumed that the planet is impartially grey and reflects the same proportion of radiation in every part of the spectrum, which is far from correct, but does not introduce any very serious error, and, further, that the planet is airless, which does, as we shall see presently, make quite a difference. But one thing is true: since the planet is in overall thermal equilibrium, it must emit the same amount of energy as it absorbs, so that its radiation as measured by an outside observer should correspond fairly closely to its greysphere temperature.

If we now put the bolometric albedo of Mars at 0·15, its greysphere temperature will be 216·4°K = −52·7°C. This is a very low temperature, but it is only an average for the whole globe and all the seasons of the year; the actual mean temperature of the subsolar point (with the Sun at the zenith) would still be +32°C, and more at perihelion.

Here we must again refer back to the Earth as control.

A Martian viewing our planet with astronomical instruments similar to ours would agree with Harry Wexler of the U.S. Weather Bureau that its bolometric albedo was 0·35, and might go on to compute from the Martian counterpart of the Stefan-Boltzman gimmick a greysphere temperature of

249·2°K = −23·9°C. He might then express surprise that the Earth was not encased in ice, for he could not know as we do that its actual mean annual temperature was +14·2°C and so some 38° above the greysphere value.

This is due to the *greenhouse effect* of the atmosphere, which resembles the glass of a greenhouse in being comparatively transparent to the shorter visual wavelengths which carry most of the solar energy (the peak emission is in the yellow part of the spectrum, for the Sun is a yellow star), but opaque to the low infra-red or 'obscure heat' emitted by the ground and objects at room temperatures. Indeed, we have already seen (p. 77) that only within the water window, between 8 and 13μ, can this radiation escape directly to space.

To compute the greenhouse effect of the Martian 'air' with any degree of assurance would require knowledge beyond our present reach. But the three atmospheric constituents that are the most effective agents of trapping the heat of the Sun in our air are water vapour, carbon dioxide and ozone, and all three are present on Mars. It is true that the amount of 'precipitable water vapour' above its surface is comparable to that of our deserts, but most of its atmospheric water exists in the form of fine ice crystals in the violet layer, and as stated (p. 69) this, too, has a considerable greenhouse effect. Furthermore, while the equivalent amount of CO_2 in our air is only 220 cm, the atmosphere of Mars is credited with as much as 9000 cm of this gas, which will cause a correspondingly stronger absorption of obscure heat. The content of O_3 may be no more than 0·1 of ours. Yet it has been calculated that its heating effect on the lower atmosphere through the absorption of the higher ultra-violet will be comparable to that of CO_2 absorbing the near infra-red.

In fact, the whole problem of atmospheric heating by CO_2, O_2 and O_3 has been mathematically investigated in some detail by C. Prabhakara and J. S. Hogan, Jr. in the U.S.A. (1965) for a range of surface barometric pressures from 10 to 50 mb. They have found *inter alia* that the tropopause of Mars should lie at about 10 km (say 6 miles) and have a temperature of about −100°C (I obtained the same result in

1959), continuing to drop gradually with further ascent down to about −120°C, whereafter the atmosphere becomes isothermal (stratosphere). These results are in fair agreement with those of other investigators.

Nevertheless, while such papers make impressive academic exercises, their practical value is doubtful. The most important point left unconsidered in the calculations of greenhouse effect is the contribution to it by mists and hazes, frequent on Mars, and their relative incidence by day, when they would cut down sunshine on the way in, and by night, when they would hinder the obscure heat on the way out.

We have seen in Chapter 7 that all observations concur at the finding that haze and cloud increase with a drop in temperature, and with the coming of the night and winter in particular. It is well known from daily experience that cloudy or misty nights are warm, whereas clear nights are cold, so that this order of things on Mars should serve to temper the daily and seasonal extremes of temperature.

The moment is now ripe to examine the observational methods of determining the surface temperatures of Mars and their results; but it must not be naïvely assumed that the temperature of the surface can be measured as such – it can only be *inferred* from the measurements of the energy of the low infra-red and/or microwave radiation emitted by the planet.

To begin with, most of the planetary heat is blanketed off by the water vapour in our air, and so long as the observations are made from the ground they must necessarily be confined to the water window. A wider range can be covered from a stratospheric balloon, but no comprehensive investigation of this kind appears to have been made so far. The peak emission of energy at the mean temperature of the Earth falls at about 11μ and so near the middle of the water window, but its wavelength increases as the temperature drops and vice versa. The wavelength of maximum emission is simply related to the absolute temperature of a black-body emitter by Wien's Law, so that the black-body temperature can be obtained from the spectral position of the maximum. It can then be corrected for the albedo. But things become difficult if the maximum lies beyond the water window.

The main awkwardness, however, stems from the fact that the measured radiation must pass on the way through the planet's atmosphere, where some of it is absorbed and re-emitted. Should haze or mist intervene the absorption may be appreciable, and in an extreme case what is measured may be the temperature of the atmosphere and not of the ground. Since a thin atmosphere such as that of Mars is generally much colder than the ground, the infra-red temperatures of the planet will tend to come out too low.

Thus W. W. Coblentz at Lowell has found that the true surface temperatures of Mars may be 5–20°C above the thermocouple readings, and G. de Vaucouleurs notes that, while the reading for the polar regions may be −40°C if clouds are present, it will rise to −20°C if the Martian sky is clear. The intervention of a blue cloud may depress the observed temperature to −60°C. The lowest temperature actually recorded on Mars is −85°C and refers to a blue cloud. Yet even if no visible clouds are there the atmosphere will intervene through the carbon-dioxide and ozone absorption and affect the result.

W. J. Humphreys gives in *Physics of the Air* (1940) a simple formula for calculating the mean temperature of the Earth's surface from the atmospheric radiation balance. If corrected for the ground albedo, taken to be 0·10 on the average (p. 65), the formula gives exactly the right figure (Firsoff, 1959). Encouraged by this result, I have tried to apply the same method to Mars.

Assuming a hazy, strongly-scattering atmosphere, which transmits directly to the ground 50% of the incident sunlight and absorbs 80% of the obscure heat, re-radiating one half of this equally up and down, I have obtained a mean annual temperature of 245·5°C = −27·5°C. The absorbed and transmitted or re-radiated fractions may be varied within certain limits. Thus if the atmosphere is taken to be thinner and clearer, absorbing 50% of the planetary infra-red, it will also be more transparent to direct sunlight, the absorbed fraction being 20%. The resulting mean temperature will be −46·5°C annual, −35·7°C at perihelion, and −55·4°C at aphelion (Firsoff, 1969). Comparison with the theoretical greysphere temperature of −52·7°C gives a greenhouse effect of barely 6·2°C, which goes to show that this alternative must be close to the possible rock bottom. Mars should be warmer than that.

Microwave measurements could be more reliable at low temperatures than the infra-red determinations owing to their higher 'signal level' and comparative indifference to mist and cloud. The published results, however, are saddled with large uncertainties. In the January 1966 issue of *Planetary and Space Science* R. D. Davies and D. Williams have reported that the microwave observations of Mars at 21·2 cm in 1963 and 1965 give a 'brightness temperature' (i.e. black-body temperature calculated on the assumption that the measured radiation is wholly thermal in origin) of 271 ± 76°K. Radio waves of this length can penetrate a thickness of rock up to about 2 feet, so that they would normally refer to the subsoil of Mars, but the wide margin of error robs the result of practical value. Earlier observations at shorter wavelengths gave temperatures near freezing point. C. W. Allen (1963) quotes a microwave figure of 220°K, which is very close to the greysphere temperature. The Russian radio astronomers Kutuza, Losovskii and Salamonovich obtained in 1966 a mean brightness temperature of the disc equal to 225 ± 10°K. To sum up, for the time being the infra-red temperatures look much more reliable.

During the period from 1924 to 1932, covering five oppositions, Edison Pettit at Mt Wilson made systematic thermocouple scans of Mars, which gave the following average temperatures:

	Subsolar Point	Limb	South Polar Cap
Perihelion	300°K (+27°C)	279°K (+6°C)	222°K (−51°C)
Aphelion	273°K (0°C)	254°K (−19°C)	201°K (−72°C)

It will be noticed that the subsolar temperature falls considerably below the theoretical grey-body mean of +32°C, although temperatures of up to 35°C have been recorded. An atmospheric effect lowering the reading may be involved.

More recently W. M. Sinton mapped the diurnal temperature changes at two points of the Martian surface on 20 July 1954. These are tabulated below with reference to the local Martian time, the day of Mars being divided into 24 hours:

Latitude	Martian Time (in hours)							
	7:00	8:00	9:00	10:00	11:00	12:00	13:000	14:00
+10°	−64	−41	−19	−2	13	26	28	15°C
−2°	−64	−32	−16	−1	16	22	22	7°C

The midday temperatures in the two tables are in fair agreement, but the discrepancy between the morning readings by Sinton and the limb temperatures obtained by Pettit is sharp. This may be partly due to Pettit's having taken a mean for the morning and evening terminator (limb), whereas Sinton gives only the morning temperatures. Yet it could hardly account for the whole of the difference, and thin blue or yellow clouds seem to have contributed to the effect.

It will be recalled from the earlier discussion that both the violet layer and the ground haze thicken in the late afternoon and both terminators look misty (Fig. 12). Overnight frost is to be expected on Mars, and isolated white patches, interpreted as hoar-frost, possibly on hilltops or crater rims, have been seen near the morning terminator even in the equatorial zone. A temperature of −64°C, however, should result in a general frosting all along the terminator, and none such has been reported. Moreover, a glow of vaporisation has been observed there. This has been examined by Henri Camichel with the Lyot polarimeter at Pic du Midi and found to display the polarising properties of water droplets about 2μ in diameter (1959). Such droplets occur at low temperatures in the summits of our cumulus clouds and can exist in supercooled condition down to the so-called Schaeffer point of

−39°C, at which water vapour crystallises spontaneously in the absence of condensation nuclei. Yet −39°C is a long way above −64°C, and even at 9 o'clock Martian time the temperature is well below freezing. Now, while liquid water can be supercooled without freezing, it cannot arise in the first place without the temperature having risen above freezing point. This appears to show that Sinton's morning temperatures are spurious and refer to a high atmospheric level. We have seen on page 88 that some atmospheric correction would be necessary in any case.

No such correction appears to have been made when W. M. Sinton and J. R. Strong summarised (1960) their findings for perihelic conditions in the equatorial zone in putting the maximum midday temperature at 25°C in the deserts and 33°C in the maria, with a rather uncertain night minimum of −75°C. Another estimate (Mintz) makes the polar minimum −103°C. N. A. Weil (1965) estimates the diurnal temperature range at 97°C, and goes on to conclude that 'the maximum air temperature in the near-surface (2-metre elevation) zone' will be 55°C lower than the observed surface maximum, which yields a temperature of −23°C at the 'base of the troposphere'. This is once more difficult to reconcile with a fog of water droplets along the terminator, and, while such reasoning embodies an element of truth, it seems to lead to a greatly exaggerated picture of the situation, owing to the all-too-literal, mechanical acceptance of the undigested laboratory and instrumental data.

One of the difficulties is that we have no observations of the dark hemisphere of Mars, but even more important is the uncertainty of the ground barometric pressure, which has been put variously at anything between 5 and 136 mb! At the time of writing the low values are much in vogue on the strength of the interpretations given to the *Mariner* 'occultation experiment', but, as has been repeatedly pointed out, this result derives from speculative assumptions and appears to be wholly incompatible with the bulk of observational material, or even with the appearance of the *Mariner* photographs (p. 49). A value of 50 mb will, therefore, be provisionally assumed as a compromise. (1 mb = 0·75 torr or mmHg, and 50 mb = 37·5 torr.)

From the range and relative rapidity of temperature changes recorded on the surface of Mars it has been inferred that the soil of Mars, especially in the desert areas, must be devoid of a significant moisture content, and its very low conductivity and thermal inertia require the surface to be covered with fine dust. Victor Kachur (1967) writes, 'Firsoff (1959) has emphasised the importance of gas sorption effects in describing the lunar environment. His presentation is directly applicable to Mars, where a CO_2-rich atmosphere is expected to undergo a physico-chemical coupling with the planet's dusty surface.' He proceeds to show from the diurnal temperature variation that a correlation appears to exist between the conductivity or the Martian soil and that of a CO_2 atmosphere, indicating energetic sorption of CO_2 by the former.

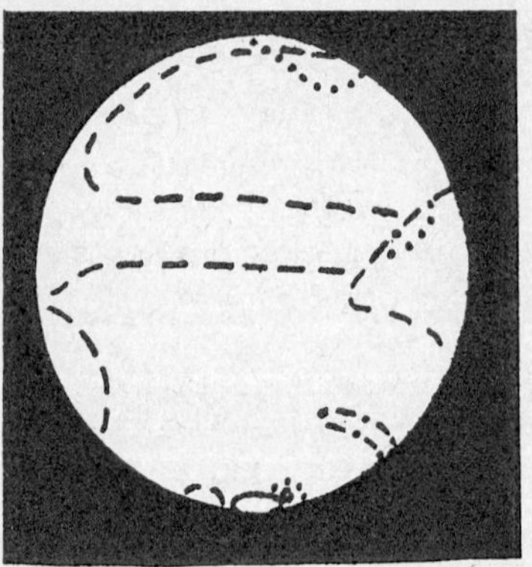

Fig. 12. *Observational drawing by C. F. Capen, 16 May 1967, 03.15 U.T., using a 24-inch reflector. - - - - - - cloud outline, ... white frost area.*

Very similar arguments were put forward in the case of the Moon, but the conception of a dust-covered Moon has not survived the close-up views supplied by the surface probes; these have shown instead that the surface rocks have a loose porous structure. It has been proposed here on the general geological grounds (p. 21) that Martian volcanics will be of a similar texture, more particularly in the areologically recent parts of the surface. This, however, does not affect the situation so far as gas sorption is concerned, as such porous rocks are also excellent sorbents. Since, though, they must be assumed to be predominantly siliceous and silicates have a sorptive affinity to water rather than CO_2, the content of sorbed water vapour in at least some parts of the Martian surface should be relatively high. On the other hand, carbonic acid should be a dominant factor in chemical weathering, which would result in the formation of carbonates alongside of silica, and the former would have a sorptive preference for CO_2. Such processes would be particularly important if the barometric pressures are low, but they would operate under the assumed 50 mb as well and may be ecologically significant. It seems, indeed, permissible to think that the difference between the maria and the terrae is related to them.

Yet we have seen that the reality of the temperature changes is in some doubt, because, while the high midday temperatures measured when the sky of Mars is clear may correspond fairly closely to the actual condition of the surface, the low figures assigned to the terminator most probably refer to the atmosphere. This may be an issue of more-or-less rather than either-or, but the resulting difference could be crucial to such Martian life as there may be.

However we may interpret the position, it remains undeniable that in the absence of seas the climate of Mars must be of an extreme continental type, resembling that of our deserts, where night frost is not infrequent even at the equator. It is none-the-less misleading to base the assessment of the Martian situation, as is often done, solely on the instantaneous amount of precipitable water vapour in the atmosphere. Two factors must be borne in mind here: the low barometric pressure and the low temperature of the atmosphere. This may not be 55°C below that of the underlying surface within six feet from it, but it will certainly drop to −100°C or less by the time the tropopause is reached, so that any water vapour ascending to it must be substantially frozen out. The pressure of water vapour in equilibrium with ice at −50°C is 0·03 torr, dropping to 0·008 torr at −60°C, to become negligibly small at −100°C, the assumed temperature at the top of the troposphere. Now 0·008 torr corresponds under Martian gravity to 0·029 g/cm^2 of precipitable water vapour, which lies a little above Suslov's mean figure of 0·020 and Dollfus's 0·015 g/cm^2, and would give a dew point of about −65°C. This, however, is only an average for the whole of the visible surface of the planet and for the whole of its atmospheric depth: there will be more water vapour locally and at lower levels, which may be partly screened off by mist or haze, and certainly less at higher levels. The important point is that at the top of the troposphere (tropopause) and over the polar regions in winter the saturation point will be readily reached and passed, and, once this has happened, no amount of increase in the quantity of ice crystals or surface deposit of hoar-frost, ice or snow, whichever it may be, will affect the vapour content.

The condition of equilibrium between the solid, liquid and vapour phase of H_2O is a highly complex meteorological function of temperature and barometric pressure.

For pure water to occur, however evanescently, in liquid phase in the open the barometric pressure must be at least 4·58 torr. Capillary water and briny pools could exist at lower pressures, but would be subject to very strong evaporation.

If, though, the ground atmospheric pressure is 6 mb, which equals 4·50 torr, water would boil without melting, or *sublime*, at about −½°C, which completely rules out any 2-micron water droplets along the sunrise terminator, let alone a blue line at the edge of the retreating polar caps (p. 30). This does not ring true any more than an atmosphere of pure carbon dioxide it implies.

At the provisionally assumed ground barometric pressure of 50 mb, or 37·5 torr, water will boil at 32°C, and we have seen that such temperatures are reached on a perihelic summer afternoon in the tropical maria. I suspect that the true pressure is higher, more like the 83 mb originally proposed by Dollfus, which would raise the boiling point to 42°C. But this does not make any great difference in the present context. On any reckoning the rate of evaporation will be very high, so that liquid water could not survive for long anywhere on Mars, except possibly, as stated, in briny pools.

It follows from this discussion that the main peculiarity of the Martian climate is the very rapid evaporation of surface water, except such fraction thereof as may be held by sorption in the porous or powdery materials, the resulting vapour being whisked off by the ascending currents into the intensely-cold high atmosphere, where it is frozen out in the violet layer and eventually precipitated at the poles. Thus, although the instantaneous content of precipitable water vapour in the atmosphere may be low, this provides no measure of the total amount of water as substance on the surface of Mars. Water is cosmically a very common compound (p. 20) and, quite apart from any underground waters (p. 55), the amount of water in the violet layer and other hazes and clouds and especially in the polar caps must be a high multiple of that disclosed by the spectroscope.

This régime immediately explains the vital importance of the polar caps to the observed seasonal cycle.

Comparisons with the Earth are very useful, provided we do not forget the physical disparity of the two planets. It is not enough to raise the surface of the Earth to 50 000 or 80 000 ft above sea-level to obtain Martian conditions. The gravity on Mars being 2/5 of ours, 2½ times as much air will be needed to produce the same pressure. The input of solar energy is between 1/3 and 1/2 of the terrestrial. The atmospheric composition is different. The seasons are about twice as long. Thus the two situations are in no way comparable.

In particular the latent heats of evaporation and freezing involved in the atmospheric circulation provide an efficient mechanism of heat transfer from the surface to the atmosphere, and if the absorption of the infra-red radiation from the ground by CO_2 and O_3, as well as of the solar ultra-violet by the latter, is likewise taken into account, the lower atmosphere of Mars could not be anything like as forbiddingly cold

(−23°C has been suggested – see p. 91) as is often assumed. Of course, Martian nights and winters are cold. Yet summer temperatures of about +10°C have been registered even at the south pole of the planet, and there can be little doubt that the usual picture of Mars is badly overpainted. No reliable numerical estimates are possible in the absence of adequate data.

There is in Tibet an isolated plateau of about the same size as the State of California, having an average elevation of 15 000 ft and surrounded by still higher mountains. It is called Chang Tang. Few travellers have visited it, and our information on its physical condition, flora and fauna is limited, but it may form, as has been suggested by G. P. Schilling (1962), a fair approximation to the Martian situation.

Most of the area is bare rock and soil, but grasses and low plants occur here and there. They do not grow to over 3 inches and send deep tap roots into the soil. The daily variation of temperature is very wide, the minimum occurring shortly before sunrise and the maximum soon after noon. Owing to the very low atmospheric water content, there is no precipitation through most of the year, and only seldom do light cirrus clouds veil the sky. Strong ascending air currents form in the morning, though there is no record of any 'glow of vaporisation'. The temperature seems to stay constant above 1·5–2 km over the ground and the jet stream passes near the surface of the land, which – Schilling seems to think – may be an instructive portrayal of what happens on Mars. This I doubt, for the reasons stated.

Still, Chang Tang is not devoid of life, modest as this may be.

9. Within the 'Ring of Life'

The so-called 'Ring of Life', within which the heat of the Sun is neither excessive nor inadequate to sustain the 'carbohydrate' biochemistry evolved on the Earth, extends from somewhere inwards of the orbit of Venus to a little beyond that of Mars. This means that, other things being favourable, life similar to ours could exist on the Red Planet. There is, of course, 'many a slip twixt the cup and the lip', and it could turn out that 'other things' were not favourable – there could be too little water, or not enough air, and so on. On the other hand, we are not too well placed to judge what forms life may or may not take, or what conditions would make it impossible, because our experience is limited to the terrestrial example, where organisms have become adapted over a period of 3000 million years or so to precisely this particular planetary environment, and correspondingly misadapted for survival in other possible environments, such as the Martian.

Nearly 370 years have gone since Giordano Bruno was burned alive in Rome for the heinous crime of believing in the plurality of inhabited worlds. The distance covered by the human mind in the intervening period must be measured in more than the mere passage of time. Our knowledge has increased and so has our tolerance. Yet the basic outlook which sent Bruno to the stake lingers on under different labels.

It is a matter of daily experience that any suggestion, any finding favourable to life on our celestial neighbours is greeted with a frown of professional disapproval, unless perhaps it be qualified by such life being 'lowly'. Intelligent life somewhere inaccessibly deep in hypothetical space is a different matter – it does not affect the 'here and now' and leaves Man's position as the Lord of Creation substantially unchallenged.

Contrariwise, most improbable ideas and fanciful hypotheses may be sure of inaudible applause from a large section of astronomical opinion if they lead to the conclusion that there is only one life-bearing planet in the Solar System – the Earth.

This is supposed to be the hallmark of 'scientific attitude', embodied in the distinction, often drawn, between 'natural' and 'biological' explanations, with special reference to Mars. Yet what scientific principles can this outlook appeal to?

It has long been recognised that the Earth cannot occupy a privileged position in the universe, or for that matter in the Solar System. Uniformity of nature would lead us to expect life to be widespread and varied. Alternative biochemical schemes are certainly conceivable, even probable, so that life could flourish in surroundings deadly to our kind.

Thus there should be parity of esteem between organic and inorganic explanations: there is no intrinsic reason why the latter are to be preferred. Even intelligent life exceeding human achievement cannot be ruled out. All that can be said against it is that it should not be invoked wantonly as a kind of supernatural agency – a *deus ex machina* – where a simple answer eludes us. Nothing should be accepted uncritically, but, alas, much is on the mechanistic side of the argument.

The general appearance and distribution of the maria, the latter of which favouring the south tropical zone where the summer is warmer and avoiding the corresponding northern latitudes with their cold aphelic summer, as well as of the minor dark surface features associated with them has been described in Chapter 3. We have seen that they wax and darken, often experiencing striking colour changes, with the coming of the Martian spring; and, conversely, wane and pall as the polar caps begin to expand beneath a canopy of mist and cloud. Some seemingly freakish changes in colour or appearance have also been noted, as well as permanent or long-term alterations of extent or outline. And the curious thing is that wherever the latter type of change has occurred it has invariably resulted in augmenting the total dark area.

The relationship between the seasonal behaviour of the dark features (henceforth referred to as maria for the sake of simplicity) and of the polar caps is clear and unmistakable, which becomes easy to understand if we reflect (p. 95) that nearly all of the surface supply of water – or perhaps to be accurate, that active part of it which is directly involved in circulation – is contained in the latter. It is, therefore, obvious that in some way or other the shade and colour of the maria, perhaps their very existence, depend on the moisture released in the retreat of the polar caps. That this is not a matter of rising temperature alone is clearly shown by such facts as that Lowell's 'wave of quickening' does not progress polewards from the tropics, where the Sun stands highest in the sky, but the other way round, and that it overruns the equator into the opposite hemisphere by as much as 1000 miles.

An outside observer would see a similar 'wave of quickening' on the Earth in the annual revival of vegetation in spring in the temperate zones. But this 'wave of quickening' moves the opposite way, following the retreating snows towards the poles. Thus on the Earth this effect is primarily temperature-dependent, for, as we know full well, there is no shortage of surface water, and this is not whisked off the ground by fierce evaporation, to be imprisoned in a high crystal velum (though a *cirrovelum* as such is not unknown, especially in winter) and eventually precipitated about the poles. The Martian economy is the direct result of low gravity, combined with low atmospheric pressure.

This is a clear reason why, if it is assumed that the maria are vegetated areas, their tide of life should follow an opposite seasonal course.

Their total area exceeds 10 million square miles. If all of it were overgrown with vascular plants, which account for most of the Earth's vegetation, absorptions of chlorophyll should appear in the spectra of the maria. Yet a diligent search for these by G. P. Kuiper has failed; all he could find was a slight strengthening of the telluric water bands, indicating that there was more water in or above the maria than in the deserts.

The explanation, suggested by James Franck (1949), was that Martian plants could use some other type of photosynthesising pigment, as exemplified by our bacterial chlorophyll, which may assume various colorations, but is typically purple. This could also account for the rather odd colour changes observed in the higher latitudes, where the previously greenish areas tended to turn brown, maroon, purple or even carmine with the coming of the spring (p. 31). The point that the oxygen liberated in bacterial photosynthesis is not released into the air, but used to oxidise other substances, chiefly hydrogen-donor molecules with water as one of the products, could be a useful adaptation to Martian conditions, as well as serve to explain the apparent scarcity of oxygen in the atmosphere.

This sounds reasonable, but the argument is somewhat circular, for, given the high abundance of CO_2, a Martian flora ought to have experienced no difficulty in liberating sufficient oxygen to modify the atmospheric composition in the course of areological time, as green plants are assumed to have done on the Earth. In any case this type of speculation is too Earth-bound to be really useful. One of the first adaptations Martian plants may be expected to develop, to parry the wide amplitude of temperature fluctuations between day and night and the danger of desiccation, would be thick or hairy skins, which would make them largely non-transparent, thus hiding both their water content and chlorophyll from the gaze of the spectroscope. This would also account for the predominant greyness of the maria as compared with our vegetation.

Indeed, the reflection curves of the maria, obtained by Kuiper, differed sharply from those of our grasses or deciduous trees, though they could be matched by our lichens and dry mosses, which on simple reasoning would be best fitted to survive on Mars. It would, however, be taking too much for granted to assume that Martian plants must necessarily be of this type. In fact, in experiments with growing plants in simulated Martian conditions (S. M. Siegel, G. Renwick, O. Daly *et al.*, 1965) it was not mosses, ferns or liverworts, but 'higher plants', such as cucumber, cockscomb, millet and rye, that have done best. It may be that these experiments will have to be remade if the ground-level barometric pressure proves lower than was assumed; but we must await further evidence before accepting the conclusions drawn from the 'occultation experiment'.

In any event a sealed glass jar cannot adequately deputise for a planet, so that such experiments, however interesting, can be only permissive but not prohibitive; in other words, success in raising and maintaining terrestrial organisms in a jar environment may be taken to indicate that they could survive if transferred to Mars, but failure would have no conclusive power as regards possible Martian life.

The changes in the protein structure and metabolism of plants grown in quasi-Martian habitats, where they began to liberate such unusual gases as CO and H_2, may serve as a further indication that the chemistry of indigenous Martian floras would be different from that of terrestrial vegetation, even if based on the same general plan.

A further lion in the path of the vegetative interpretation of the maria was their infra-red appearance. Our plants look almost snow-white when photographed in infra-red, whereas the dark areas of Mars show 'coal-black' through an infra-red filter.

This was one of the reasons why alternative inorganic explanations have been preferred by some. The earliest of these is due to the Swedish chemist Svante Arrhenius, author of the hypothesis of *panspermia*, according to which dehydrated spores were propelled through space by light pressure and acted as germs of life when finding a suitable world. Arrhenius proposed explaining the colour changes observed in the maria by assuming that these were areas covered with incrustations of hydroscopic salts, which altered colour when wetted. This explanation, however, created more problems than it solved, and is no longer very popular.

One of the main objections to any purely mineralogical

interpretation of the maria, apart from the difficulty of satisfactorily reproducing their colour behaviour, has been voiced by E. J. Öpik. On an arid world, such as Mars, dust must be formed by rock decay, volcanic action and possibly meteoric infall, and the winds would spread this dust more or less evenly over the surface, concealing in an areologically short time any intrinsic differences of colour. To withstand this age-long onslaught, the colouring matter requires regenerative powers, such as those of growing plants.

It is, nevertheless, true that exposed ridges could escape being buried under a pall of dust, and it has recently been suggested by Carl Sagan and others that the maria, far from being depressed regions, like those of the Moon, are, on the contrary, elevated mountain land, which is swept clean of dust by the prevailing winds. This interpretation appears to be based on the roughness of some maria indicated by radar echoes (p. 54). Since, though, radar makes use of radio waves in the centimetre range, the roughness recorded by it must be of the same order, and could easily be due to small bushes or even clumps of grass. Besides, the tip of Syrtis Major, which is typical marial terrain, is radar-smooth. The photographs obtained by *Mariner 4* (Plates XIII–XVI) are quite inconclusive in this respect: they only skim the edge of Mare Sirenum, where, it is true, numerous craters appear. But craters appear in lunar maria as well, while still bigger craters and higher mountains are shown by frames 11–14 in the desert regions of Atlantis and Phaetontis. The argument does not click, but even if it were shown, as it has not been, that the maria of Mars were uplands and not lowlands, this could not by itself account for their darkening, let alone changing colour, with the coming of spring.

Sagan's hypothesis lacks any explanatory power.

The same objection applies to D. B. MacLaughlin's idea that the maria owe their colouring to volcanic ash spread by the winds (1955). Volcanic ash could, no doubt, be 'regenerated' by further volcanic eruptions and so escape being buried under dust; it could be redeposited by the winds in the same approximate pattern; it could spread to new regions.

But it is very hard to see how and why it could alter in colour to brown, purple or carmine when the nearest polar cap begins to shrink. Moreover, the shapes and borders of the maria are often too definite to be due to so haphazard a cause; some other explanation would be required for even the incontestable 'canals' and oases, which appear to contain moisture withal (p. 38).

Nevertheless, there is something in MacLaughlin's suggestion, for the pattern of dark features on Mars appears to be related to that of prevailing winds. Volcanic matter may be involved in this, too; but, if so, from what is known about the general behaviour of the maria the decisive factor would probably be water emitted by volcanic vents. Indeed, on the Earth it accounts for about 90% of all volcanic exhalations, and, since underground water is expected on Mars (p. 55), the same should be true of Martian vulcanism.

According to yet another hypothesis, proposed by the Kiesses and Karrer (p. 79) in 1960, everything was supposed to be due to oxides of nitrogen, appearing in solid, liquid and vapour phases, NO_2 being variously balanced against N_2O_4 depending on the temperature and pressure, with smaller amounts of NO and N_2O_3 as transitional forms. The caps were supposed to consist of frozen N_2O_4, which is chalky-white, but would become yellowish with rising temperature and partial conversion to NO_2 (this would require melting, which is impossible under Martian barometric pressure). Some of the colour changes and Kozyrev's 'green haze' could be accounted for on this basis.

Yet the required range of temperatures falls far below those of Mars. Thus N_2O has a boiling point of −118·3°C at 40 torr, N_2O_4 would sublime without melting at −14·7°C even under 100 torr and at −23·9°C at 40 torr. The material of the caps has been shown to be water ice. Further spectroscopic work has failed to confirm any absorptions due to nitrogen oxides, for which rather low limits have had to be fixed (p. 81), such limits being significant, as the concentration of these gases in our air is negligibly small and there is no overlap with telluric lines.

The nitrogen-oxide hypothesis was, of course, meant to put an end to 'nonsense' about Martian life, as these compounds are poisonous to most terrestrial organisms. Oddly enough, however, dinitrogen tetroxide (N_2O_4) is an excellent water-like solvent, freezing at −11·2°C, and as such could provide a basis for pseudo-organic chemistry on a moderately cold world. You may dismiss this as airy speculation, but some down-to-earth beetles include nitrogen oxides in their metabolism. It is an order of 'respectable antiquity', as N. W. Pirie puts it, and he suggests that such 'activities that we now look upon as odd . . . may once have been widespread' (1954). So you never know.

The latest inorganic arrival (1966) is the hypothesis put forward by S. Otterman and F. E. Bronner, also in the U.S.A., attributing the 'wave of darkening' to 'soil frost phenomena'. It is argued that the alternate thawing of the ground by day and freezing by night, involving the moisture from the cap, produces 'minute, frost-heaved, soil surface features', which give the ground a porous structure, yielding a darkening effect. This cannot happen unless and until day-time temperatures rise above freezing point.

While such structures are not impossible, they could not result in colour changes, the 'wave of darkening' would progress from the equator to the pole, and the whole idea appears to be rather far-fetched. Moreover, the very low atmospheric pressure and the corresponding low water content of the soil that have been lately canvassed would make such an explanation even more difficult to sustain than the simple recourse to vegetation.

The inspiration has probably come from Dollfus's finding (1961) that the polarisation curve of any given marial region maintains the same general form through the seasons, but the proportion of polarised light rises during the spring and summer. Laboratory tests appear to indicate that the light is scattered by small objects, some 0·01 cm across, whose diameter and numbers vary in an annual cycle. This has led Carl Sagan to suggest that the maria may be covered with micro-organisms or small inorganic particles responsive to moisture. Yet similar polarisation curves could, no doubt, be produced by hairs or other growths on large plants, so that our understanding of the situation is not greatly advanced by this information.

To sum up, with all due respect to J. W. Salisbury (*Icarus*, Vol. 5, No. 3, 1966), none of the inorganic hypotheses so far proposed offers nearly as natural and satisfactory an explanation of the seasonal behaviour of the maria as does the simple annual cycle of the growth and decay of a vegetative cover of some kind or other. I would, on the other hand, strongly endorse his view that access to underground water may be the main factor that differentiates the maria from the deserts – in fact, I have said so myself.

We have seen in Chapters 7 and 8 that it is a mistake to think of Martian temperatures and pressures as direct instrumental data: they are only *deduced* from observational measurements and the processes of deduction involve a sizable element of speculation. The margins of uncertainty may not be large by stellar standards, where 100°C this way or that is of no account, but they are large enough for a delicately balanced ecological whole, such as a planetary environment. It is, therefore, unwise to reject simple visual and photographic observations of the planet, because they do not seem to fit one or another set of theoretical figures.

At this point we may advantageously return to the dark appearance of the maria in 'infra-red light'.

For some time after World War II this problem was investigated in detail by the late G. A. Tikhov and his 'astrobotanical' team at the Academy of Sciences of the Kazakh Republic, in which they could draw on the material from the land-locked high mountain area of the Pamirs and the Siberian arctic.

The upshot of this investigation has been that the behaviour of chlorophyll varies with temperature and is important for controlling the body heat of green plants. At moderate temperatures chlorophyll strongly reflects the solar infra-red. But it does more than that: it *fluoresces*, re-emitting some of the absorbed radiation of shorter wavelengths in the lower infra-red and red register. Both these actions increase with rising temperature, thus serving to rid the plant tissues of excess heat.

According to Tikhov *et al.*, in the pine the fluorescent radiation increases 40-fold between −40 and +20°C, though it is still present at the lower limit. At the same time, however, less and less of the solar infra-red is reflected as the temperature drops, which overbalances the loss of energy by fluorescence and helps to conserve body heat. At low temperatures the plants no longer look bright in infra-red photographs. Indeed, the effect extends beyond the period of winter cold, at any rate in extreme continental climates, such as that of Siberia – and of Mars? Tikhov writes, 'In April the dispersion of infra-red rays in all trees drops abruptly; evidently at the beginning of a new period in the plant's life it stands especially in need of a considerable amount of warmth.'

This, too, seems applicable to Mars, where vegetation may require additional energy after the frosty night. On the other hand, the perihelic midsummer temperatures at tropical noon in the maria average 25–35°C and could hardly be described as low. Be this as it may, the argument from the infra-red appearance of the maria against their vegetative character, or if only against the presence of chlorophyll, loses its apparent power.

Martian plants could, of course, have developed other means to protect themselves against the extremes of temperature they have to face, and we have seen (p. 88) that the very low temperatures (down to −75°C) inferred for the night minimum may refer to the atmosphere and not the surface of Mars.

In any case, in the experiments with simulated Martian environments referred to on p. 101 an 'equatorial cycle' was used where the temperature was maintained at 20–25°C for 8 hours and at between −20 and −30°C for 16 hours, combined with condensable atmospheric water content of 0·01–0·05 g/cm^2 – and ordinary terrestrial plant seedlings survived and grew. Moreover, plants grown in an anaerobic atmosphere (no oxygen) stood up well to hard frost, while those raised in a terrestrial atmospheric mixture died.

The real diurnal temperature range on Mars is not likely to exceed these experimental conditions, and in the state of nature all changes will be far more gradual. On the Earth in the state of nature Don Juan Pond in Antarctica sustains a modest but varied microflora at temperatures down to −23°C and perhaps lower (S. A. Zernow). This is a brackish water basin, whose freezing point is estimated at −45°C. Similar pools could exist on Mars.

Living cells in cold climates can manufacture natural antifreezes, such as glycerol, other polyhydric alcohols and hydroxylamine. Owing to this the common round-leaved wintergreen (*Pyrola rotundifolia*) stays unfrozen down to −32°C. The scurvy grass (*Coclearia arctica*) which grows on northern shores of Siberia suffers no harm from exposure to −46°C (Tikhov *et al.*). R. W. Salt in Canada has investigated the chemistry of hibernating insects and found that they develop a high percentage of glycerol with the onset of winter, being thus able to remain supercooled down to −40°C. Lichens on the Arctic rocks can live in environments where temperature never rises above −5°C and continue to assimilate carbon, if at a very slow rate, down to −35°C (Tobler).

It was shown as early as 1905 by Paul Becquerel that dehydrated algae, lichens, mosses, bacterial spores and such animalculae as rotifers and tardigrades, found among the Arctic lichens, can be exposed to temperatures a fraction of a degree above the Absolute Zero of −273·1°C and to hard vacuum for long periods without harm.

Warm-blooded animals can keep their body temperature at a constant level by regulating their metabolism, and owing to this device Emperor penguins can incubate eggs in Antarctica at −60°C. It must further be remembered that the main hazards of these terrestrial cold regions come from the wind, snow and ice, not just low temperatures. These, too, are decisive, rather than low temperature or low atmospheric pressure, in the 'upper biotic zone' (18 000–20 000 ft) of the Himalayas, which yet supports a variety of plants – grasses, sedges, composites, as well as vertebrate and invertebrate animals – reptiles, bees, flies, spiders and birds, rodents and carnivores as transient visitors.

It must be repeated that we do not know for certain how much oxygen there is on Mars, but there must be some, if only from the photo-dissociation of carbon dioxide and water vapour (p. 79), and if there are photosynthesising plants these, too, ought to produce some oxygen, even if most of it is retained within their bodies. In any case, green plants do not require oxygen for survival; what they need is carbon dioxide, which is much more plentiful on Mars than on the Earth. Anaerobic animals are possible, although they would have to rely on the relatively inefficient process of fermentation instead of oxidation for their supply of energy. On the other hand, all energy requirements would be reduced under low gravity.

Some experiments have been made to test the ability of animals to stand up to a reduction in the partial pressure of oxygen. The frogs do not appear to be seriously disturbed until the total pressure drops to about 100 torr, say, 25 torr partial oxygen pressure. Bees and wasps can fly at a partial pressure of 35 torr or so under the overall barometric pressure of 1 atmosphere (760 torr). This ability should increase in inverse proportion to gravity, so that on rough reckoning about 15 torr of partial oxygen pressure should suffice on Mars. The gain, however, is negatived by the fact that 2·68 as much oxygen would be needed to yield the same pressure. On the other hand, an increase in the area of the lungs or tracheae should present no insuperable biological difficulty. And we are still thinking in purely terrestrial terms – there may be other processes, unfamiliar to our biology, of generating the necessary energy.

In sum total this evidence is inconclusive, but not unfavourable.

There exist, however, some spectroscopic data that may have a bearing on the problem.

In 1956 W. M. Sinton discovered three absorptions at 3·43, 3·56 and 3·67μ in the infra-red spectrum of Mars, which he was able to confirm in 1958, using the 200-inch Hale reflector.

He found these absorptions substantially limited to the maria, although they were faintly traceable in some desert regions, such as Arabia and Amazonis, as well. They show a good coincidence with those of terrestrial vegetation, though the third line is present only in some blue algae, and characterise the carbon–hydrogen bond in organic molecules. It has transpired, however (O. G. Red, B. T. O'Leary and W. M. Sinton, 1965), that at least in the case of the two latter absorptions confusion may be involved with the 3·58 and 3·69μ bands of telluric heavy water DHO, as a correlation appears to exist between their strength and the atmospheric water content.

Further, J. S. Shirk, W. A. Hazeltine and G. C. Pimental announced (1965) that 'the infra-red absorption bands observed by Sinton at 2710, 2793 and 2898 cm^{-1}, in the spectrum of Mars may be due to the vapour of the heavy waters, D_2O and DHO, in its atmosphere, where they have become concentrated by the gravitational fractionation of heavy isotopes'.

In combination these two findings would indicate an intervention by heavy water. On the other hand, Sinton's original interpretation finds strong support in the recent results of multiplex interferometric spectroscopy (p. 76 f.)

In the absence of confusing telluric lines this method permits to winkle out minute amounts of unusual atmospheric constituents from the infra-red spectra of the planets. Some of these constituents may be so short-lived that they could not maintain themselves in the atmosphere for long unless continuously or periodically replenished from an active source, which in the case of some organic compounds could be identified as biological.

Thus the spectra obtained during the 1964 apparition contain a number of unexpected absorptions in the 6000 cm^{-1} region, which is characteristic of the first overtone of the CH band and so a common location of substituted methanes. The spacing of the bands also points to their being due to hydrides, and at least some of the absorptions show good coincidence with known spectra. Further spectra were expected with improved equipment during the 1967 opposition, but nothing has been published at the time of writing.

Once more we have spectra characteristic of the CH bond. The concentrations of the individual compounds are estimated at about 0·1%. 'By terrestrial standards' – writes L. D. Kaplan – 'such concentrations are extremely large; for Mars they are startling.' He goes on to say that this knocks the ground from under the argument of P. Abelson and others that life could never have originated on Mars owing to the shortage of hydrogen, which the planet would have been unable to retain owing to its low mass, hydrogen compounds being regarded as necessary for the initial abiogenic syntheses of organic compounds (see *Life, Mind and Galaxies* in the same series). The issue is, therefore, changed 'from whether the Martian atmosphere can contain hydrogen to why it does, and why the reduced compounds are not oxidised in the presence of large amounts of carbon dioxide', which will be photo-dissociated to oxygen and CO (p. 79) in the condition of energetic upward convection in the atmosphere.

It is, of course, true, as Kaplan says, that substituted methanes *might* be produced by inorganic processes and so be due to 'natural' causes, but given parity of esteem for biological and abiotic explanations, there is no reason to prefer this interpretation, the more so as no such causes have yet been plausibly suggested.

The case is not proven, and must remain so until we have close-up photographs of Martian life by surface probes or some equally unambiguous evidence, but on the totality of observational data amassed so far the presumption is strong that the maria are vegetated areas.

10. The Queer Little Moons of Mars

Jonathan Swift is said to have completed the manuscript of *Gulliver's Travels* in or about 1720, but delayed its publication till 1726, because he was apprehensive about the reception of this political satire. He need not have worried: it was found 'immensely diverting'. Among the various imaginary places visited by Gulliver was the flying island of Laputa, whose astronomers had made some interesting discoveries.

> 'They have likewise discovered two lesser Stars, or *Satellites* which revolve about Mars; whereof the innermost is distant from the Center of the primary Planet exactly three of his diameters, and the outermost five; the former revolves in the space of ten Hours, and the latter in Twenty-one and a Half . . .' (pp. 170–71, *Gulliver's Travels.* Blackwell, Oxford, 1965).

Voltaire's *Micromégas*, published in 1750, also refers to two satellites of Mars, but without particulars, in which the chief interest of the case lies, for it was widely believed at the time that, since the Earth had one moon and Jupiter four known companions, Mars must have two.

The satellites of Mars had been searched for diligently for a long time without success, until they were eventually spotted in 1877 by Asaph Hall with the 26-inch refractor of the U.S. Naval Observatory in Washington and named *Phobos* (Fear) and *Deimos* (Terror) after the companions of the God of War Ares. Calculations showed that Swift had 'exaggerated as usual': Phobos revolved at only 1·38 diameters from the primary's centre in 7 h 39 min 26·6 s, and Deimos at 3·46 diameters in 30 h 21 min 15·7 s. Still, his was not a bad guess (see Fig. 13), especially if the extreme astronomical improbability of such two bodies is taken into account.

Their visual magnitudes at the mean opposition distance are 11·6 and 12·8, and they would have been over a magnitude brighter during the very favourable apparition of 1877. Thus they were not all that difficult to see, except for the proximity

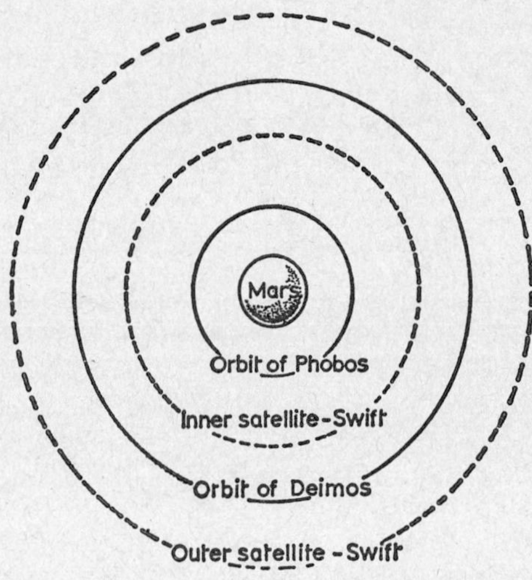

Fig. 13. *Orbits of the Martian satellites – actual in solid lines, and according to Swift in broken lines.*

of the comparatively large and brilliant disc of Mars. The apparent mean-opposition diameter of the latter is 17·87″, while the corresponding maximum angular separation from its centre is 24·6″ for Phobos and 1′ 1·8″ for Deimos, so that they are fairly drowned in the glare of the planet. It is, nevertheless, a matter for surprise that they had eluded Sir William Herschel's 48-inch reflector and Lord Rosse's 72-inch 'Leviathan of Parsonstown', completed in 1845, whose optical quality was sufficiently high to show the spiral structure of the Andromeda Nebula and some other galaxies.

In this connection a Japanese friend, Takeshi Sato, has drawn my attention to the fact, reported in *Sky and Telescope*, that both satellites were observed from Japan in 1956 with quite modest apertures. Thus Deimos was seen by H. Araki with a 6-inch (15-cm) reflector and by several other observers with 8-inch telescopes, while Saheki spotted Phobos with an 8-inch altazimuth reflector.

Phobos is the larger, as well as the closer of the two. Its orbit is inclined at 1° 6′ to the equatorial plane of the primary, and with an eccentricity of 0·0210 is only some 2% out of true circle. Its mean semidiameter is 5826 miles, so that the little moon cruises over the Martian equator at an average height of 3715 miles, giving rather better views of the planet than *Mariner 4*; but, as K. A. Ehricke has pointed out, it would make an unhandy landing place for Earthlings, because entering an orbit in the equatorial plane of Mars from an approach trajectory substantially in its orbital plane is an expensive space manoeuvre. On the other hand, the satellite is ideally placed for a spacecraft starting from Mars.

Be this as it may, when the orbital period of Phobos is compared with the Martian day of 24 h 37 min and 22·69 s, it will be seen that, contrary to the rules of 'decent' celestial conduct, the satellite rises in the west and sets in the east. It will further be seen that in one revolution Phobos outstrips Mars by 16 h 57 min and 56·4 s. In order to obtain the Martian synodic period of Phobos, or the time in which it moves full circle in the sky of Mars, from full to full, or from new to new, in terms of phase, we must divide the day of Mars into the latter time and multiply the result into the sidereal period of Phobos. With four-figure logarithmic accuracy this yields 11 h 2 min and 6 s. Obviously, Phobos will need half this time to describe an arc of 180° in the sky as referred to the centre of the Martian globe, but the satellite is so close to the observer on the surface of Mars, that things will look rather different to him and the situation will depend on his areographical latitude.

If we neglect the inappreciable inclination and eccentricity of Phobos's orbit, the unobstructed mouse eye-view from a point *O* on the Martian equator is shown in Fig. 14, where *R* is the radius of Mars, and *D* the semidiameter (radius) of the orbit of Phobos. It will be seen at a glance that to an observer thus placed Phobos will have moved through 180°

in the sky, and so from horizon to horizon, in the time required by it to describe the angle of 2α about the centre of Mars. A simple calculation shows that $2\alpha = 137°32'$. This corresponds to 4 h 20 min and 10 s (approx.). As seen from

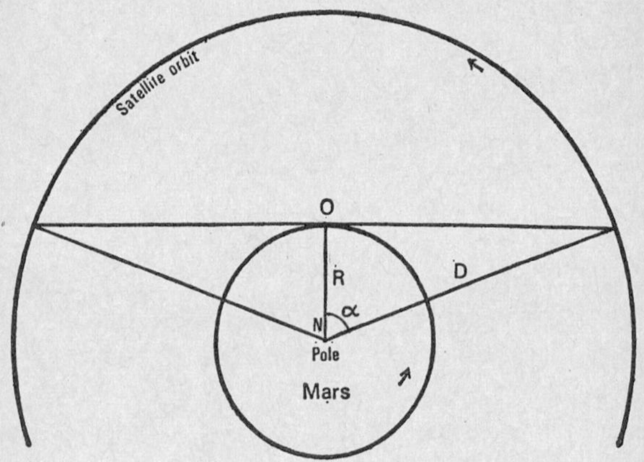

Fig. 14. *The path of a satellite in the sky of Mars as seen from the planet's equator.* (*See text.*)

higher latitudes, Phobos will move obliquely in the sky, so describing a longer arc and staying up longer, but always less than half its Martian synodic period. It may, further, be noticed that the time from the rising to the setting is appreciably over a third of the latter period, and the phase of the moonlet will undergo a corresponding change during that time; there will likewise be an appreciable difference between the phase as seen and that referred to the centre of revolution (i.e. of Mars), the two coinciding only at full and new.

The orbit of Deimos, 14 600 miles in semidiameter and 0·0028 in eccentricity, is a very nearly perfect circle and forms an angle of 1° 42′ with the planet's equatorial plane. The period of revolution of 1 d 6 h 21 min 15·7 s of our time is longer than the day of Mars, so that the second moon rises 'properly' in the east and sets in the west, but it creeps sluggishly through the sky. A calculation similar to that made for Phobos gives a Martian synodic period of 132 h 33 min and 20 s. $2a = 163°22'$, which corresponds to 62 h 3 min and 50 s.

The diameters of the two moonlets are too small for direct measurement; they are merely inferred from their magnitudes on the questionable assumption that they reflect the same proportion of sunlight as does the surface of Mars. If darker, they will exceed the now-quoted figures of 8 and 5 miles; if brighter (which is less likely, as small airless bodies are usually dark and porous), they will be smaller.

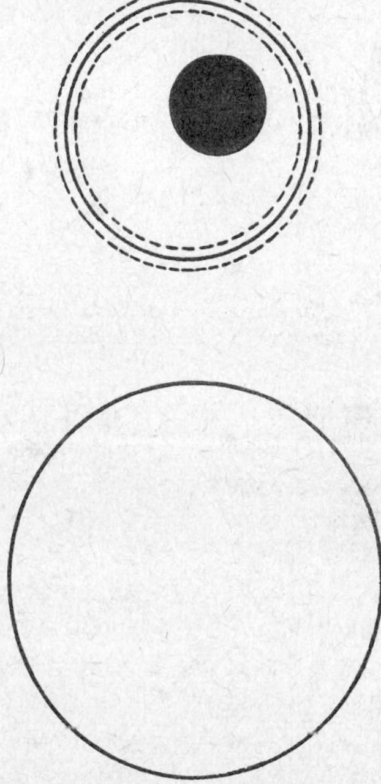

Fig. 15. *Apparent size of the Sun as seen from Mars* (top) *and the Earth* (bottom). *The solid line at the top shows the mean diameter (21′) of the Martian Sun, the broken lines the aphelic (19′ 12″) and perihelic (23′) diameters respectively. The black circle represents the satellite Phobos in transit as it might appear from the equatorial region of Mars. The terrestrial diameter of the Sun averages 32′.*

From the equatorial region of Mars Phobos will look like a little moon, some 8′ across, or a quarter of the apparent diameter of our relatively distant and splendid companion world, when near the zenith, but will noticeably decline in size towards the horizon. It will pass through all the usual phases, but its full phase will be a comparatively rare sight, as it will be eclipsed for long hours every night, except for a few weeks about the solstices. The Sun as seen from Mars (Fig. 15) subtends an average angle of 21′, so that it can never be eclipsed by Phobos, let alone Deimos, which will have an apparent zenithal diameter of about 1′ 12″ and compare in brightness to our Venus.

Deimos will miss the cone of shadow cast by Mars, except for some time in the spring and autumn, but it will need a keen sight to follow its phases. Both satellites will occasionally transit the Sun, which will make an interesting astronomical phenomenon to watch.

The moonlets are so close to the planet that they can never be seen from its circumpolar regions, Phobos from latitudes of over ±69°42′ and Deimos of over ±83°25′ (see Fig. 16). These figures are over-precise inasmuch as they refer to the *astronomical horizon*, which is the great circle traced on the celestial sphere by the horizontal plane passing through the point of observation. The real, horizon, however, lies somewhat below the astronomical by an angle known as *dip of horizon*, which depends on the curvature of the globe and the observer's elevation above its surface. On Mars the dip amounts to about 2° for the summit of a 7000-foot mountain. Atmospheric refraction should also raise the satellites above the theoretical boundary; on the other hand, they may be extinguished by atmospheric obscuration.

In sum total the situation is unprecedented in the Solar System, and the midget moons of Mars are a dynamical teaser.

There are two conceivable ways in which a planet can have acquired a satellite: by being formed together with it from the same primordial condensation (tidal fission is nowadays out of favour); by capturing an independent body, such as an

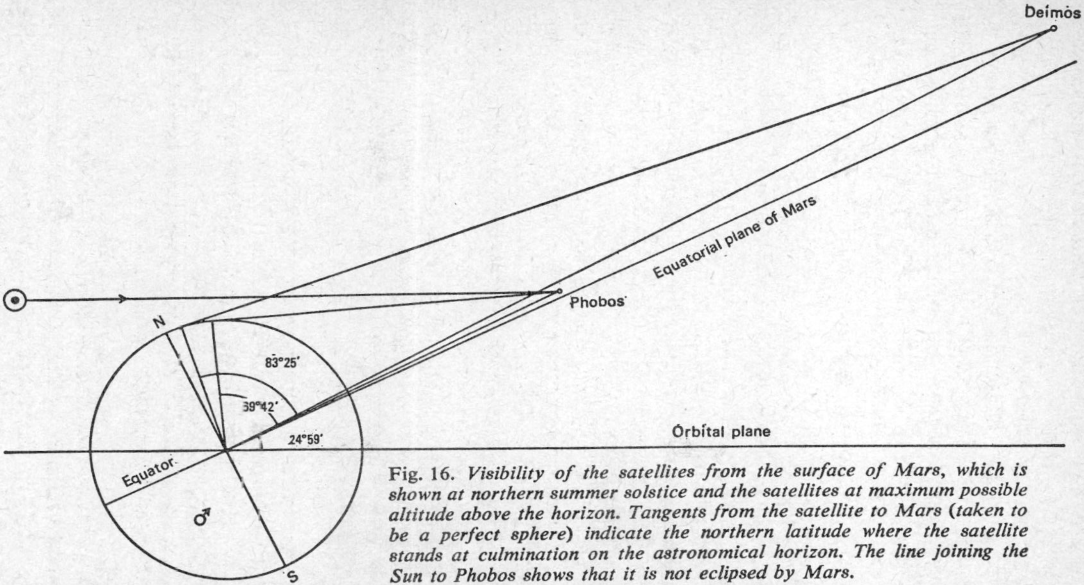

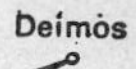

Fig. 16. *Visibility of the satellites from the surface of Mars, which is shown at northern summer solstice and the satellites at maximum possible altitude above the horizon. Tangents from the satellite to Mars (taken to be a perfect sphere) indicate the northern latitude where the satellite stands at culmination on the astronomical horizon. The line joining the Sun to Phobos shows that it is not eclipsed by Mars.*

(N.B. The figure of 24°59′ was quoted from the 1968 *Handbook of the British Astronomical Association*, and was later found to be incorrect. It should be 23°59′, as given elsewhere in this book.)

asteroid, through cumulative perturbations. The inner moons of the giant planets fall into the first category, some of their small external satellites into the second.

Phobos and Deimos could not have originated by the first process: they are far too small and too close to the primary and would have been engulfed in the mass of the protoplanet at the start. Moreover, as we shall see presently, there exist clear indications that at least Phobos could not have accompanied Mars for very long on astronomical reckoning, which would again preclude their simultaneous origin.

Yet the moonlets orbit Mars substantially in the equatorial plane, which is typical of the inner satellites, and unknown among the outer ones.

It must here be borne in mind that the equator of Mars stands at an angle of 23° 59′ to the plane of its orbit, and the most likely mechanism of capture would place the orbital planes of the acquired asteroids close to the latter and not the former. The improbability of two captured asteroids orbiting the primary in near-coincidence with its equatorial plane is so high as to amount to virtual impossibility. Moreover, captures result in orbits of high eccentricity, which could evolve to circularity only after a prolonged period of perturbation or friction in a resisting medium.

The inclinations of the orbits of the asteroidal satellites to the equatorial plane in the systems of the giant planets range from 27·7° to 163° (an inclination of over 90° results in retrograde rotation). Their eccentricities are correspondingly large, attaining 0·75 in the case of Neptune's Nereid. In addition, these orbits are wide with long periods. Thus none of the giants seems to have been able to emulate the feat of the puny Mars.

To cap it all, it has been calculated by B. P. Sharpless (1945) and confirmed by F. J. Kerr and F. L. Whipple (1954) that Phobos is approaching Mars at a rate that is high by astronomical standards and should crash on to it in between 35 and 40 million years. This also implies that it could not have been circling Mars for any geologically or areologically long period – and certainly not *ab initio*.

The cause of the secular acceleration of Phobos must be sought in the drag exerted on it by the high exosphere of Mars.

Owing to its lower mass, Mars cannot hold its atmosphere down to its surface as tightly as the Earth, and, other things being equal, gravity alone would make the main part of the Martian atmosphere extend upwards 2·68 times as far as ours. Yet the molecular spray of the exosphere is governed not by gravity as such, which declines inversely to the square of the distance, but by the gravitational potential, which has a longer reach. Thus even at 3715 miles above the surface there may be enough gas to slow down the motion of a satellite. This gas, however, will be in the constant danger of being swept away by the solar wind, so that its density should vary in the sunspot cycle.

Now the well-known Russian astrophysicist I. S. Shklovskii made some computations, and in 1959 published two articles in *Komsomol'skaya Pravda* (The Truth of the Young Communist League), in which he maintained that the density of this gas was far too low to cause the observed secular acceleration of Phobos if it were a solid body, but would be adequate if it were a hollow sphere. Since there was no known astronomical process that could generate a hollow structure of this kind, the only possible alternative was that Phobos was an artificial satellite, which would presumably go for the smaller Deimos as well. The reason why these satellites had not been discovered until 1877 was simply that they were not there to discover, having been launched shortly before that date.

Thus either Mars was a home of superior technological civilisation, as assumed in Lowell's canal hypothesis, or it was used as a base by the operators of the 'flying saucers'. This sounds like science fiction of the reddest kind, but was a perfectly justifiable conclusion if the premises were granted.

Komsomol'skaya Pravda is not held in any great respect as a scientific publication, but Shklovskii is a serious scientist, so that his ideas have not been simply laughed out of court. The position was re-examined in 1963 and 1964 by G. F.

Schilling in the U.S.A. with the conclusion that some 'reasonable assumptions' regarding the physical state of the equatorial exosphere of Mars lead to a density adequate to exert the assumed drag on Phobos even if it is not a hollow sphere. He goes on to state that his 'theoretical investigation can neither prove that drag accelerations in the mean daily motion of Phobos do exist nor disprove that the satellites of Mars are of artificial origin'.

His assumptions, however, include a ground barometric pressure of between 83 and 133 mb, which is much above anything accepted today, and a highly hypothetical concentration of interplanetary meteoric dust in the gravitational field of Mars.* He also mentions the possibility of Phobos being a ball of ice, which seems to be more than questionable, as in this case it could not survive very long at Mars's distance from the Sun. In fact, the albedo of Phobos is more likely to be comparable to that of the Moon than Mars (small asteroids have very low albedoes), which would increase its cross-sectional area by about 2·7 and its diameter by $\sqrt{2{\cdot}7} = 1{\cdot}64$ to over 13 miles. If it is a ball of rock its density could not be much less than 2, and, to judge by the meteorites, assumed to be of asteroidal origin, could be much higher. To sum up, while the issue is certainly open, it cannot be disposed of as easily as that.

The midget moons of Mars are very queer astronomical bodies even without Shklovskii.

* *Mariner 4* has registered an increase in the flux of interplanetary dust particles by a factor of 5 outwards beyond the Earth, the dust being densest between the planets, but no increase in the vicinity of Mars itself (W. M. Alexander and C. W. McCracken, 1965).

11. The Mystery Remains

Discoveries are made and unmade by further discoveries. What was thought final yesterday is discarded today as untenable, and it needs no great acumen to predict the same fate for some of the opinions now in fashion. This is not to deny all progress. The techniques are improving all the time. There is an accumulation of observational data. But our quest is difficult, and the latest findings are not necessarily also the most reliable. In a process of continual growth there is room for all information, irrespective of date.

Although the evidence is strong that some kind of life exists on Mars, the verdict must remain 'not proven' for the time being. The question of an advanced order of intelligent life is more speculative, but it is too early – and indeed unwise – to jump to the conclusion that it is impossible.

Certainly, *Mariner 4* has shown no 'canals', no artificial structures. Yet if 'canals' were pipelines they would not be visible as such, and, the photographs having been taken during a Martian winter, the strips of vegetation supposed to extend along them would not be recorded either. The same would be true if these followed areological fractures exuding life-sustaining moisture – at least the fractures are there.

In an attempt to assess the value of the *Mariner* evidence with regard to life, S. D. Kilston, R. R. Drummond and C. Sagan have examined (1966) thousands of photographs of the Earth, obtained by the *Tiros* and *Nimbus* meteorological satellites and showing the surface with a resolution of the order of a few tenths of a kilometre. None of them, however, affords any unmistakable sign of life, with the possible exception of the regular grid of forest fellings, underlined by a fresh fall of snow, near Cochrane in Canada, in a picture taken by *Tiros 2* on 4 April 1961. Even here, though, our interpretation would be far less certain if we did not know in advance what we were looking at. Similar, quite regular grids can be discerned in the *Orbiter* views of the Moon and some of the *Mariner* frames (Plates XIII–XVI), but both are, no doubt, the effect of 'geological' stresses.

A 'streakiness' (described in connection with the latter) intersecting the scanning grid will be seen in some satellite photographs of the Earth as well, and in this case is quite certainly atmospheric. This may have an application to Mars, and in any event goes to show that linear atmospheric features can arise without jet aircraft, not to mention meteorites. This is pertinent to the case of the putative jet vapour trail in one of the *Nimbus 1* frames. As for the interstate highway described by the authors of the report in another *Nimbus 1* photograph, showing Tennessee, I am unable to make it out at all in the printed reproduction of the original, and would undertake to show quite a few similar lines in the *Mariner* views of Mars. The straight 'breakwater' off the Moroccan coast appearing in another *Nimbus* frame looks highly artificial, but is, in fact, a natural feature, partly submerged.

The resolution of *Nimbus 1* pictures is about ½ km, say ⅓ of a mile, while the smallest compact features, such as craters, visible in the best *Mariner* frames are some 3 miles across, and their definition is much worse owing to the facsimile telegraphic technique employed in this case, whereas the weather-satellite views have been transmitted by television. It is well-known, however, that elongated features, such as fractures or, for that matter, interstate highways, can be seen even when well below the resolution limit in width. Thus the Tennessee Interstate Highway is not more than 50 metres (0·05 km) wide.

A closer approach to Mars by a photographic satellite, with transmission distance cut down to an opposition minimum to give better definition, could possibly reveal artificial structures, if they exist. With a satellite placed in an orbit about Mars photographs could be taken in batches at intervals over a long period, to study seasonal changes attributable to vegetation. Yet Kilston *et al.* remark, 'The remote detection of seasonal variation in the reflectivity of vegetation is difficult to perform because of the general low contrast between the vegetation and the underlying terrain.' Satellite photographs of the Earth's argicultural areas show no trace of a checkboard of crops. On the other hand, a *Nimbus* frame of the Fergana Basin, in the Kirghiz Soviet Socialist Republic, contains a dark area bearing a vague resemblance to Tithonius Lacus on Mars. The latter planet at the eyepiece seems to be more obliging so far as seasonal changes are concerned.

The presence of an atmosphere, be it as thin as it may, is a complicating factor in the probe exploration of Mars, as it makes close orbiting impracticable. An orbital satellite could not photograph the planet from 28 miles, as *Orbiter 2* did the Moon. Nor could a *Ranger*-type craft be successfully employed, for it would be destroyed by atmospheric friction long before reaching the surface, unless appropriately slowed down by counter-jets. Whether a surface probe could be landed by parachute is at present in doubt, owing to the uncertain value of the density of Martian air; but there is nothing to prevent using the lunar method of rocket control, or perhaps some combination of both.

A surface probe could, of course, reveal large organisms or artificial constructions by direct photography. The total area of Mars, however, is 55 million square miles, some 75–80% of which is classed as desert, so that, unless the target area can be very accurately pinpointed, the unpredictable element of luck must enter into the result.

In any life system micro-organisms must be much more numerous and ubiquitous than large forms, so that the chances of detecting life increase on the microscopic scale. With this in view, although – I suspect – also from the unreasoned conviction that any life beyond the Earth must be 'lowly', various methods and devices have been proposed and developed to deal with the problem at this level.

One such device is the *Wolf Trap* (W. Vishniac, 1960), which is designed to admit external material upon landing on a planet and pass it on to a container with nutrient fluid. As the trapped micro-organisms multiply the fluid decomes increasingly turbid, which is registered by means of a scanning beam of light and the result communicated to Earth by radio. The *Gulliver*, developed by G. V. Levin *et al.* (1962), shoots out little projectiles attached to the apparatus by sticky tapes, to trap microbial life. These are then retracted into the probe for study. Once more a nutrient fluid is involved, and, as Francis Jackson remarks, 'There is no reason to suppose that Martian organisms will necessarily be able to grow in a mixture of compounds that would support terrestrial organisms.' This rather limits the usefulness of this and similar devices, although they have been successfully tested in terrestrial conditions. Tricky problems of sterilisation and control likewise arise.

J. Lederberg has suggested instead using an automatic microscope with a television camera for the direct study of the acquired planetary material. Gas chromotography and other means of chemical analysis are likewise possible. Martian soil could be dug up and scooped into the apparatus in the same way as has been done on the Moon by the American and Russian probes.

Owing, however, to the greater distance of Mars, all this requires much more powerful transmitters, and simple close-up photography is easier and may prove to be more effective than the relatively elaborate biological contraptions. In any event they will probably be preceded by a photographic reconnaissance.

As the book goes to press (March 1969), one *Mariner* is already on the way to Mars and another is to follow shortly. A closer approach with a flyby taking the probes over both poles is intended. There are also rumours of the Russians planning to attempt a soft landing on Mars, so that our understanding of the situation may be greatly improved in the near future.

We can but wait and see.

Bibliography

ALLEN, C. W. 1963. *Astrophysical quantities*, 2nd edition. Athlone Press, London.

BRITISH ASTRONOMICAL ASSOCIATION. 1967, *Handbook*.

EHRICKE, K. A 1962. *Space Flight, Vol. II, Dynamics*. Van Nostrand, London.

FIRSOFF, V. A. 1959. On the radiation balance in the atmosphere of the Earth, Venus and Mars. *Journal of the British Astronomical Association*, **69** (5).

FIRSOFF, V. A. 1963. *Life beyond the Earth*. Hutchinson Scientific Publications, London; Basic Books, New York, 1964.

FIRSOFF, V. A. 1964. *Exploring the planets*. Sidgwick & Jackson, London; A. S. Barnes, Cranbury, N. J., U.S.A. Revised edition – in the press.

HODGMAN, C. D. (ed. in chief). 1962. *Handbook of chemistry and physics*, 4th edition. The Chemical Rubber Publishing Co., Cleveland, Ohio, U.S.A.

JACKSON, F. and MOORE, P. 1965. *Life on Mars*. Routledge & Kegan Paul, London.

KILSTON, S. D., DRUMMOND, R. R. and SAGAN, C. 1966. A search for life on Earth at kilometer resolution. *Icarus*, **5** (1).

KUIPER, G. P. and MIDDLEHURST, B. (eds.). 1961. *The Solar System*. Vol. III. University of Chicago Press.

LOVELOCK, J. E., HITCHCOCK, D. R. *et al.* (symposium). 1967. Detecting planetary life from Earth. *Science Journal*, **3** (4).

LOWELL, P. 1909. *Mars as the abode of life*. New York.

MAMIKUNIAN, GREGG and BRIGGS, M. (eds.). 1965. *Current aspects of exobiology*. Pergamon Press, Oxford.

RICHARDSON, R. S. 1954. *Man and the planets*. Muller, London.

SLIPHER, E. C. 1965. *Mars – the photographic study*. Sky Publications, Cambridge, Mass., U.S.A.

STRUGHOLD, H. 1954. *The green and red planet*. Sidgwick & Jackson, London.

TIKHOV, G. A. 1955. Is life possible on other planets? *Journal of the British Astronomical Association*, **65** (5).

VAUCOULEURS, G. de. 1954. *Physics of the planet Mars*. Faber, London.

WEIL, N. A. 1965. *Lunar and planetary surface conditions*. Academic Press, London and New York.

Index

Other titles in this series
from Oliver and Boyd
are described on the following pages

CSP 1

BASIC ASTRONOMY

Patrick Moore

It is an astonishing fact that although Aristarchus of Samos was seriously suggesting that the Earth moved round the Sun, his ideas did not prevail over the Ptolemaic System and it was left to Copernicus to spark off the great astronomical awakening of the Renaissance.

In this book Patrick Moore traces briefly the history of astronomy, developing his narrative into a full account of Earth and sky, the Sun, the Solar System, the stars, the galaxies, and quasi-stellar objects, or quasars. Such knowledge does not come from theorising alone, but from innumerable observations through the agencies of the naked eye, optical and radio telescopes, and the spectroscope. The author describes these instruments together with the principles of their use.

It is also the business of astronomy to speculate about the origins and evolution of the universe. We are therefore given a glimpse of the various theories that have been proposed from the Abbé Lemaître's notion of the primeval atom in the early 1920s to present-day speculations. We are left with the thought that it is surely the height of conceit to suppose that our particular Sun is unique in being attended by a life-bearing planet.

Patrick Moore is well known to the general public through his other books and his BBC television programme *The Sky at Night*. He is Director of the Armagh Planetarium in Northern Ireland.

CSP 2

LIFE, MIND AND GALAXIES

V. A. Firsoff

First and foremost an astronomer, Val Axel Firsoff also has a vast knowledge of experimental facts in biology, chemistry, and biochemistry. With a fine intuition on the nature of mind in relation to the rest of physical reality he is able to look on the natural world with a breadth of vision that is rare nowadays.

In this book he reflects on the origins of life, examines its chemical basis, and illustrates the possibility of alternative biochemistries for life elsewhere in the universe. The warp and weft of his cosmological fabric are living things and the universe, both of which are shown to evolve towards higher patterns of organisation. In each case experimental evidence supports the argument.

Perhaps the most controversial part of this work is the contention that mind is an essential part of physical reality, which leads the author to discuss the possible relationship between subatomic physics, consciousness, and free will.

V. A. Firsoff is the author of numerous scientific papers and has lectured widely to university and other societies on the subject of the universe. This is his twentieth book.